Changing Rural Systems in Oman

Changing Rural Systems in Oman

The Khabura Project

Roderic W. Dutton

Routledge
Taylor & Francis Group

LONDON AND NEW YORK

First published 1999 by Kegan Paul International

2 Park Square, Milton Park, Abingdon, Oxon OX14 4RN
711 Third Avenue, New York, NY 10017, USA

Routledge is an imprint of the Taylor & Francis Group, an informa business

First issued in paperback 2016

Copyright © 1999 Roderic W. Dutton

Transferred to Digital Printing 2009

British Library Cataloguing in Publication Data
A catalogue record for this book is available from the British Library

ISBN 13: 978-0-7103-0607-4 (hbk)
ISBN 13: 978-1-138-97014-4 (pbk)

Publisher's Note
The publisher has gone to great lengths to ensure the quality of this reprint
but points out that some imperfections in the original copies may be
apparent. The publisher has made every effort to contact original copyright
holders and would welcome correspondence from those they have been
unable to trace.

To Esther, Jake and Joss who shared it all

Contents

Acknowledgements

The work discussed in this volume was initiated by Professor Howard Bowen-Jones, University of Durham. Shell International Petroleum Company generously funded the research surveys. Petroleum Development Oman equally generously sponsored and funded the Khabura Project to the end of 1985. They, with fuel contributions from Shell Markets (Oman), also gave the Project and its staff much appreciated friendly help and logistic support even beyond 1985.

Each phase of the work was approved by the Government of Oman. Relations with the Ministry of Agriculture and Fisheries were always close, leading to full sponsorship and funding by the Ministry in the period 1986 to 1994. From 1982 to 1989 the Ministry of Social Affairs and Labour sponsored the spinning and weaving projects. From these Ministries, and other branches of government, we gained many insights into the development process. Throughout the whole Project period advice and encouragement was also provided by the Diwan's Adviser on the Environment, who originally helped establish the Project when working for PDO.

The author had the privilege of being centrally involved in the Project from 1973 to 1994. During that time, specialist contributions to the research and development activities were made by more than forty senior staff, whose thoughtful hard work and dedication is fully appreciated. Three of them, seconded to the Project under the APO Scheme of the UK's ODA (now DFID), importantly contributed to partial final evaluations of the Project. Another six people, not otherwise mentioned here because their work was only loosely related to the subject of this volume, made invaluable team contributions: Jo Butler (mother and child welfare), Robert Whitcombe (research into the Little Bee, at al-Khabura), Morgan Manley and Paul Boyles (honeybees, Rustaq), Jan Karpowitcz and Alan Berkeley (honeybees, Dhofar). To all these people, and their spouses and children, many thanks.

Throughout the whole period the field work was as much a pleasure as a task thanks to the unfailing courtesy, good humour and responsiveness of the very many people with whom we interacted each day in the area of al-Khabura and other parts of rural Oman.

Tables

Figures

Acronyms

Acronym	Full name
ADAS	Agricultural Development and Advisory Service
APO(S)	Associate Professional Officer (Scheme)
ARS	Agricultural Research Centre
CCCP	Contagious caprine pleuro-pneumonia
DAP	Desert Agricultural Project
DAW	Department of Animal Wealth
DFID	Department for International Development
DM	Dry matter
DR	Death Rate
EC	Electrical Conductivity
ET	Evapo-Transpiration
FAO	Food and Agriculture Organisation
FMD	Food and mouth disease
FRC	Fibre-reinforced cement
GCC	Gulf Co-operation Council
GMDP	Goat Multiplication and Development Project
HP	Horse power
ITDG	Intermediate Technology Development Group
MAD fibre	Modified Acid Detergent, fibre
MAF	Ministry of Agriculture and Fisheries
MCI	Ministry of Commerce and Industry
ME	Metabolisable Energy
MJ	Mega Jules
MoSAL	Ministry of Social Affairs and Labour
MWR	Ministry of Water Resources
NCDP	National Community Development Programme
OBAF	Oman Bank for Agriculture and Fisheries
ODA	Overseas Development Administration
OFM	Oman Flour Mill
PAMAP	Public Authority for Marketing Agricultural Produce
PDO	Petroleum Development (Oman) Ltd
PFT	Project Field Team
PMSG	Pregnant Mare Serum Gonadotrophin
PPR	Peste de petits ruminants
PVC	Polyvinyl chloride
RDC	Rural Development Centre
RLRS	Rumais Livestock Research Station
RO	Rials Omani
SAR	Sodium Absorption Ratio

SIPC	Shell International Petroleum Company
SOAF	Sultan of Oman's Air Force
SQU	Sultan Qaboos University
UNDP	United Nations Development Programme
(WQ)ABARC	(Wadi Quriyat) Animal Breeding and Applied Research Centre

Glossary

Acacia tortilis	(plural used)	a shrubby tree
Acacia nilotica		a tree
'ammar		retainer/labourer, Batina gardens, living in garden of absentee owner
as-Sabi'		the fourth
barusti		see *da'an*
bidar	bidars	irrigator and/or date palm specialist
binus		a system by which a poor family will obtain a calf, for example, by sharing the costs of looking after the cow
da'an	du'un	sheet made from tied palm fronds
dehen		local butter
dellu	dellu	large leather container to lift water from wells, used bull power (see *zagira*)
dishdasha	dishdashas	dress-like garment worn by men
falaj	falajes/aflaj	traditional system for providing water
fardh		variety of date palm
fariq	fariqs	a *shawawi* household
gelba	gelbas	irrigation basin
ghaf		*Prosopis cineraria* - a tree
gharab		bag for holding dates, on the Batina (see *khasafa*)
gidad		method of date harvesting by cutting the whole hand
gidha'		beam made by splitting date palm trunks
gidhdha'		carpenter who works with date palm trunks
graz	graz	tool for making the walls of *gelbas*
haris	haris	guard
Id		festival
Id al-Adhha		the major religious festival of the year
Id al-Fitr		religious festival marking the end of Rahadhan
karab		butt end of the palm frond (used in *shashas*)
khalab		task of cutting the *karab* from the date palm
khallab	khallab	specialist who cuts the *karab* from the palm
khanjar	khanjars	curved ceremonial dagger, worn by men

kharaf		date harvesting by picking individual fruit
khasafa		bag for holding dates, in the Dhahira (see *gharab*)
khus	khus	palm frond leaflets
leben		butter milk
mann	mann	unit of weight - either 4kg or 5kg
mansul	mansuls	woollen rug, locally woven
mebselli		a date variety, picked immature, boiled and exported to India as a livestock feed
mikyal	mikyal	unit of volume - equivalent to 1.6kg of wheat
mingal	mingal	curved, serrated-edged knife for cutting alfalfa
miselle		needle made from a sliver of date palm frond
nebat		date palm pollen
nebat gharif		date palm pollen used to pollinate *fardh* palms
Prosopis cineraria		a tree (see *ghaf*)
qafir		general purpose basket made from plaited palm leaflets
qarata		*Acacia nilotica*
shammar	shammar	specialist who tanned hide and worked leather
shasha	shashas/shash	traditional float-boat of palm fronds
shawawi	shawawi	non-Bedouin pastoralists
semra		*Acacia tortilis*
sidra		a tree (see *Ziziphus spinachristi*)
simmat khabat		very large mat with handles, used to gather leaves knocked from *Acacia* trees
suq	suqs/aswaq	market
tahdir		tying the date hand - to a lower frond
tahzim		tying the date hand as the dates are maturing - to allow the dates to dry on the tree
tana		selling the crop unharvested, in the field or on the tree
tasgir		tying the date very low on the fronds - used for higher value interior dates
thug	thug	double donkey pannier - used to carry earth
umselli		poor-quality date found on the Batina

'uwal		dried shark
wakil	wakil	agent - sometimes appointed by people living in Hamra to look after their land in al-Araqi
wazar	wazar	wrap around cloth worn under the *dishdasha*
wilaya	wilayas	an administrative region (like county or governorate)
Ziziphus spina-christi		a tree bearing a small edible fruit (*nubuq*) - see *sidra*
zagira	zagiras	trip bucket system for raising water from wells using bull power (see *dellu*)

Project expatriate personnel

Barry Agnew, 1986-7, Project Manager
Andy Barker, 1988-9, Farm Manager
Richard Barnwell, 1981-7, Agricultural Extension Officer, then Field Manager/Extension
Graham Bell, 1976-9, Farm manager
John Stacey Birks, 1973-5, Demographic and Social Geography Research
Howard Bowen-Jones, 1972-9, Director
Adrian Brockett, 1979-80, Field Manager
Sally Brokensha, 1979-81, Spinning/Weaving
Czech Conroy, 1991-2, Economics of GMDP
Kevin Cooke, 1979-81, Agriculturalist
Gigi Crocker, 1977-8, Spinning/Weaving
David Cuthbertson, 1979, Field Manager
Mohan David, 1987-93, Handyman
William Donaldson, 1974-6, Fisheries and Marketing Research
Roderic Dutton, 1973-94, Agriculture Systems Research, then Field Director, then overall Director
Nick Foster, 1987-9, Irrigation
Andrew Gauldie, 1987-8, Veterinarian, then Field Manager/Veterinarian
Charlotte Heath, 1984-9, Spinning/Weaving
Francis Hillman, 1983-5, Irrigation systems/Appropriate Technology
Richard Hudson, 1991-2, Dairy cattle
Robin Jackson, 1989-92, Farm Manager, Wadi Quriyat then Rumais
Shafi-ud-Din Khan, 1986-94, Secretary/Accountant
Louis Kwantes, 1990-3, Veterinarian
Richard Landon, 1973-5, Soils Research
Sally Letts, 1973-5, Hydrology and Irrigation Research
Alison Lochhead, 1982-4, Spinning/Weaving
Richard Massey, 1979-81, Agriculturalist/Acting Field Manager
Fadhil al-Nakib, 1990-3, Livestock genetics
Kate Rogerson, 1988-90, Dairy, Khabura, then Rumais
Iain Rogerson, 1988-90, Extension/Training, then Field Manager
Harvey Sherwood, 1983-8, Livestock husbandry, then Field Manager
Alison Sherwood, 1988, Dairy
George Sidgwick, 1991-3, Farm Manager, Wadi Quriyat
Mike Steele, 1981-3, Agriculturalist
Mark Stephens, 1991-3, Forage and Irrigation Systems
Nick Taylor, 1990-1, Livestock productivity
Paul Ward, 1990-4, Dairy/Project Manager
Gerard Winstanley, 1981-3, Irrigation/Appropriate Technology

Changing Rural Systems in Oman

The Khabura Project

1. Introduction

Oman in the decades prior to the 1960s was largely isolated from the rest of the world and its changing economies and societies. It was very difficult for foreigners to get into the country and almost equally difficult for Omanis to return if they had departed to find work elsewhere. There was no modern education system, other than three small junior schools in Muscat. There was no health service, other than an American mission hospital, and there were no paved roads, other than a strip a few kilometres in length linking Muscat with Mutrah. International links were minimal, and the government was a small and very personal affair in the hands of the then Sultan, Said bin Taimur (Allen, 1987; Clements, 1980).

Rural communities in northern Oman had very little contact with the Sultan's government, which was based in the southern province of Dhofar. In a world in which people in most countries, including the Gulf States, gained at least some benefit from modern education and health services, Omani villagers and pastoralists had recourse only to Koranic schools and traditional healers. On the other hand, however, they retained full responsibility for the management of their rural resources on which they depended for their livelihoods and for life itself, and had evolved effective communal systems for their development and conservation. These were exemplified by regulations governing the protection of trees and by the work of the committees which controlled the traditional *falaj* water supply network. People worked interdependently, responding to the contributions made by other members of the rural communities in a system of mutual self-reliance. They also lived in harmony with their environment in a manner which time had proven to be truly sustainable.

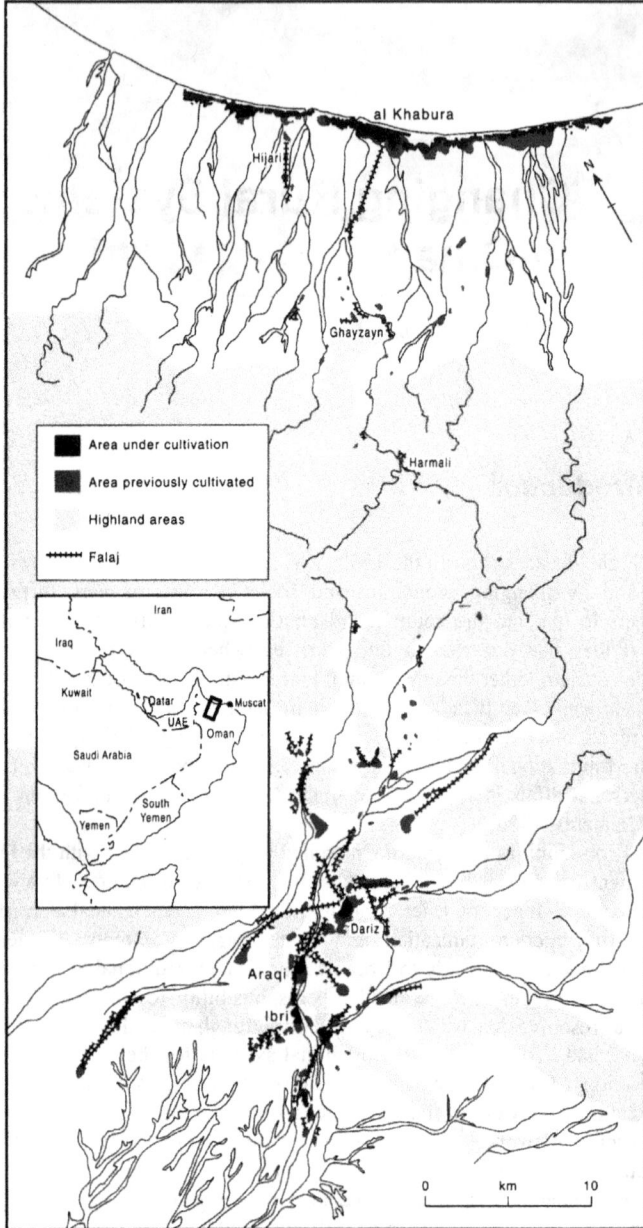

Figure 1. The research survey field study area, al-Khabura to Ibri

However, change was occurring. In the 1960s it had three main sources. First, disaffection between the Sultan and the mountain people of Dhofar led to insurrection in the mid-1960s supported from socialist South Yemen after Aden achieved its independence from Britain in 1967. This drew increased British and world attention to Oman including pressure for change that led to the replacement of Said bin Taimur by his son, the modernising Sultan Qaboos bin Said, in 1970. The second source of change was the oil enriched level of economic activity in the Gulf States which created many openings for men in rural Oman who migrated from their villages to find waged labour in Bahrain (from as early as the 1930s), Saudi Arabia and the Emirates. The material, health, educational and other aspirations of these men were deeply affected. At the same time the economic life and the organisational structures of their villages and pastoral communities were increasingly altered by their absence. The third, and most profound, source of change was the export of oil from Oman, starting in 1967. Oil income immediately became a very high proportion of national income, so its effects rapidly penetrated into every aspect of Omani rural life (Dutton, 1987a), including those which characterised even the most remote rural areas.

After Sultan Qaboos came to power in 1970, change was extremely rapid. He opened the country to external influences of all types. Within a few years, networks of schools, hospitals and roads had been built, and a conventional governmental structure had emerged, based in Muscat and the greater capital area (Skeet, 1992). Many new employment opportunities were created within the civil service and the armed forces, and attracted men into jobs away from the rural areas, though typically they continued to regard the village as their family home. These changes radically improved family incomes and standards of health and education, and the rural population grew rapidly as did its demand for water and land. People were much less dependent than previously, for life and livelihood, on the proper management of rural resources.

At the same time the new government was very keen to encourage economic growth in the rural areas. As a first step it needed to learn more about the resource base and the economic potential of the different regions. Therefore it invited a number of consulting firms to survey the country as a whole and its principal sub-divisions. In addition, the University of Durham was asked to adopt a different and complementary approach to information gathering in rural areas (Bowen-Jones, 1971). In work financially supported by Shell it undertook a series of long-term research surveys (1993-6) in a cross-section of northern Oman from the village of al-Khabura in the centre of the Batina coast to Ibri in the Dhahira (Figure 1). The research topics included a wide range of human and natural resource issues. Research was staggered but each research activity typically extended over one to two years so there was sufficient time to gain a depth of understanding about the natural resources and the people, and the nature and rate of the impacts on them being brought about by the three sources

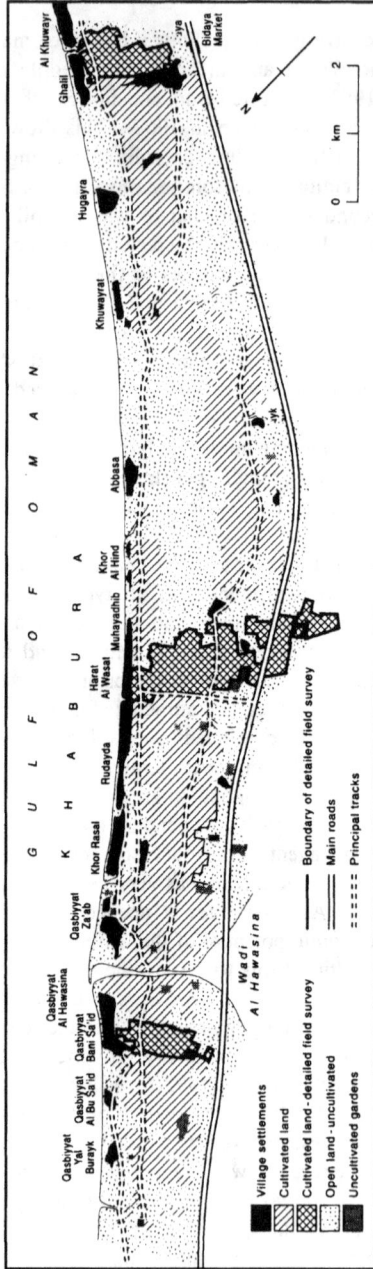

Source: Dutton, 1982b

Figure 2. al-Khabura: coastal villages and cultivated land, mid-1970s

of change identified above. The findings and conclusions of the research are detailed in chapter two of this volume, particularly where they relate to the village communities in the vicinity of al-Khabura. The University was then asked to undertake a rural development project, centred at al-Khabura (Figure 2), based on the findings and conclusions of the research (Bowen-Jones, 1974a-b). The design, implementation and outcome of what became generally known at the 'Khabura Project' are covered in chapters three and four.

In a long drawn out decision-making process initiated by the Minister of Agriculture and Fisheries in 1986 the Khabura Project changed, in 1989, from being focused around a broad-based set of farm systems and rural development activities with a sub-regional impact to having a narrower range of activities with a national scope, as discussed in chapter five. In the process it also moved from al-Khabura to new work centres in al-Kuwair (near the Ministry of Agriculture and Fisheries, MAF), Wadi Quriyat in northern interior Oman and Rumais on the Batina coast.

The Khabura Project aimed to achieve rural community development, or indeed 're-development' following the dislocation and disintegration of traditional rural systems which had resulted from the shock of the extraordinarily rapid, regional and Omani, oil-induced changes experienced in the 1960s and 1970s.

A lot was learned by the Project and real achievements were made though the overall programme fell short of its higher aims and objectives. Throughout this volume aspects of the Project's strengths and weaknesses, and successes and failures, are discussed in the hope that these evaluations will have generic relevance to similar work elsewhere. The final chapter, Prospect and Retrospect, examines the role that could be played by an entity, here called a Rural Development Centre (RDC), set up to help maximise the likelihood of achieving and sustaining beneficial change in rural areas. Finally, the Khabura Project's progress is analysed in terms of the proposed RDC structure.

The overall Khabura Project, including the research phase and the later work at Wadi Quriyat, Rumais and al-Kuwair, is most unusual in terms of its timespan (1973-94), breadth and continuity. These are certainly important reasons for any deeper insights into the working of rural systems in Oman and, more broadly, rural development processes, that the Project may have revealed. It also made the Project's staff very aware that it was working within a national political, economic and social context that was constantly changing and therefore impacting on the Project's work in different ways. The fundamental change was the rapid growth of the population and the ever increasing demand by the people for water, land, education, health, vehicles, housing and other material goods and services. Education, we saw, led to profound changes in aspirations, away from an interest in rural life, and to a reduction in environmental awareness. Also, although in the 1970s it seemed that everyone could find salaried employment in Muscat or the Gulf, by the late 1980s such opportunities were rarer. Young men, therefore, were reconsidering the option of staying in the village, and some were prepared to give more serious consideration to farming as a primary means of obtaining a livelihood. Other indicative examples of changes to which the Project had to respond include the facts that when the Project started there was no electricity in al-Khabura and no garage workshops, but long before it closed, the village was on the national electricity grid and workshops abounded.

At the national level the Project's lifespan coincided with the tenure of no fewer than three Ministers of Agriculture and Fisheries, and countless changes of other senior officials and expatriate advisers. The Ministers had strong, but contrasting, interests in the Project's work. Other officials were helpful, obstructive or passive according to circumstance and personal inclination. With some, discussion and exchanges of views were positive, with others negative. Governmental priorities, policies and regulations were also always under review, and affected the Project's work directly or indirectly. Thus an approach or an activity that might have been 'right' at the start of the Project would certainly have been less right five years on, and possibly quite inappropriate by the end. The Project's long life, therefore, made it very aware that specific objectives, approaches and techniques had always to be subject to review through constant interaction with the farmers and rural producers, on the one hand, and the national governmental machinery on the other hand.

2. Rural communities under change: development or disintegration?

The picture drawn here of rural communities has been created from the results of the series of studies, mentioned above, undertaken by a team from the University of Durham in the early and mid-1970s of a cross-section of northern Oman. The study area extended from the village of al-Khabura in the central part of the Batina coast, across the Batina coastal plain, up the wadi al-Hawasina to the pass over the Hajar mountains, and down the wadi al-Kabir to the villages of the Dhahira and the desert edge beyond them (Figure 1). The studies were made at a time of increasingly rapid change induced by growing oil wealth. However, there was also enough left of the physical, social and economic structure of the pre-oil period to envisage the rural communities as they had been before oil exports began from Oman or its neighbouring Gulf States.

In order to define the term 'rural community' and emphasise its virtues the positive characteristics of an idealised pre-oil rural community in northern Oman are itemised, taking the Dhahira village of Araqi as an example. This underlines the real strength and virtues of a rural community, as here defined, but it also shows how vulnerable to economic and social dislocation the rural community became when faced with rapid change. Pre-oil, the idealised rural community maintained its virtuous condition only by reason of its isolation from the world beyond its borders. Change, when it came, brought many benefits, in standards of health and education for example, and in some respects change meant that the small rural communities were beneficially subsumed into what was becoming a more integrated national community. Nevertheless, it will be argued here that the process of change brought many and real losses to the local communities, causing rural dislocation as an unfortunate by-product of rural and national development.

The detailed studies focused on anthropological and demographic parameters, and agriculture, hydrology and water management, soils, livestock, crafts, fisheries and marketing. They describe the rural and farm systems as they had been and illustrate the true nature and extent of the dislocation of the rural communities associated with the rapid changes that was in full flow by the 1970s. They also, therefore, indicate the important requirement at that time for a 're-development' of Oman's rural communities. The requirement, however, was not to recreate these communities as they had been in pre-oil days, which would in any case have been impossible due to the all-pervasive nature of the oil economy and the growing national role of the government. The requirement was for growth based on acceptable modifications to the range of traditional

7

activities, designed to achieve higher yet sustainable levels of economic returns whilst maintaining community structures intact.

2.1 An idealised pre-oil village community

A principal characteristic of our idealised rural community of Araqi was the interdependence of its inhabitants (Dutton, 1983b). Each member of the community played a communal role, and each member of the community was reliant upon the other members playing their roles. This inter-reliance was the essence of the concept of community, and was the glue which held the community together. Each person was critically aware of the importance of the contribution made by other individuals towards the maintenance of the community as a whole. Interdependence was consequent upon the fact that the rural community catered for most of its own needs, everyone using their complementary skills. A high proportion of food and other material requirements were either grown or manufactured locally, by making full use of local resources and by developing the necessary human skills to undertake the work involved. Non material needs, including educational and spiritual needs, were also provided by local specialists in these fields.

Interdependence created a significant degree of effective independence from the world outside. Independence shielded the idealised rural community from changes in the world that lay beyond the community's control. And to the extent that it was truly reliant on its own resources the community knew, in a fundamental way, that it alone was responsible for the maintenance of the community; responsible economically, socially, materially and spiritually. Independence, therefore, threw responsibility for maintaining the community onto its members. Accepting such a responsibility implied an awareness on the part of the community as a whole that each individual had to return as much to the community's social, material, economic and spiritual system as he or she took from it. This balance had to be maintained not only to ensure that everyone today could satisfy their rightful requirements, but also to ensure that the rightful requirements of future generations would also be catered for.

The creation and maintenance of the balance between satisfying today's requirements and providing for future needs, in turn implied a form of control to prevent the balance being lost. The control, however, had to be dynamic; it had to be flexible and responsive to changing circumstances.

One consequence of accepting responsibility for catering for the needs of future generations was that our idealised community lived in harmony with its physical environment. The farmers used water carefully, sowed their own seed, and kept their land in good heart. The artisans understood the nature and value of the raw materials with which they worked, and therefore treated them with due respect. Such a community also took responsibility for its own education, training and spiritual well-being. The idealised community as a whole may be said to have understood the meaning of 'sustainable development' long before the phrase was coined.

The idealised characteristics of the village community given above are based on a series of studies of the village of Araqi, in the Dhahira, made as part of the study of a cross-section of northern Oman mentioned above. By the time the study was being made the village was already undergoing rapid change, and it is doubtful whether the village in this idealised form ever truly existed. But it is hoped that the idealisation will usefully illustrate two points. First, the benefits that accrue to the community from interdependence and independence, and second, the vulnerability of the community, and the interdependence and independence of its people, to the forces of rapid regional change induced by oil.

Five decades ago, before the discovery of oil in Oman, or the full exploitation of oil resources in other parts of the Arabian peninsula (Penrose, 1971), it could have been argued that rural communities in Oman had retained their traditional independence based on the effective interdependence of the people living in them. The inhabitants of a village such as Araqi were mutually dependent on each others skills. With these skills they made full and effective use, at the level of technology then available to them, of the local resources. Primordial amongst these resources was water, provided for the village as a whole by a single communally-owned *falaj*, a traditional system of raising water for irrigation which had the great virtue of not being capable of draining the aquifer. To maintain the *falaj* for future usage, to work the discipline of the six day rotation of water, and to irrigate continuously night and day according to an exact time formula modified to fit different winter and summer requirements demanded a high order of responsive control and interdependence (Dutton, 1981).

In Araqi land was lowered, levelled, terraced, cultivated and manured as required to keep it in good heart. Local seed was used for successive crops of wheat and alfalfa, and local suckers for the next generation of date palms. Local animals provided the manure and a local bull the traction power for the metal-tipped plough. The plough itself was made from local timber by the village carpenter, the metal tip made by the local smith, and the guide-ropes from date palm fibre.

All buildings in Araqi were made by their owners or by a combination of local craftsmen. These men relied entirely on local raw materials: stone; a gypsum cement; mud for bricks; mud and wheat chaff for plaster; split palm trunks, knotted palm fronds and mud for roofing; and coloured mineral deposits for decoration.

Every part of the date palm was used: dates as food and the principal cash crop; date stones boiled for cattle; the palm fronds for making *barusti* sheeting (palm fronds bound together with palm fibre rope, making a strong flexible sheet of material measuring some 5m by 2.5m); palm leaves for mats and baskets; the fruiting stems as brushes; and the frond butts as fuel. Animals (camels, cattle, goats and sheep) were fed on local produce and provided

9

transport, draught power, meat, milk, skins, wool and dung, which were all fully used.

The wealth of skills required to do the above work was naturally matched by a close understanding of the constraints of the environment in which the people of Araqi lived. Crops were grown which thrived in the fierce summer heat or which could mature to harvest in the short winter cool season, and which could tolerate the system of *falaj* irrigation which brought water to any particular field only once in six or eight days. The local varieties of livestock though not very productive of either meat or milk were hardy under local conditions of management and environment.

The members of the community also lived in a state of mutual dependence with their immediate neighbours. There were employers and employees and free-lance specialists who hired out their skills, there was mutual self-help whereby, for example, farmers in turn would assist each other with the wheat harvest, there were primary and secondary producers and traders, there were craftsmen and artisans, there were village leaders and committees and there were, of course, the women who raised their families and kept domestic order as well as practising some crafts. There were also various exchanges between the village and the Bedouin who grazed camels and other livestock in the surrounding plains. The Bedouin needed dates and wheat which they obtained in exchange for their own surplus goods, including animals purchased for slaughter, dung for the land and salt.

Other goods from the village were sold in the regional market in Ibri, or taken by camel further afield. With the money were purchased those few goods that were wanted but not locally producible: rice, coffee, dried fish, cotton cloth, cooking fat, a few simple domestic supplies, and precious metals for jewellery.

The village ran its own social services. The aged and disabled were mostly looked after by their own families. The produce of particular date palms and other crops was sold to fund a charity to aid certain unfortunates and also to support the Koranic teacher who was responsible for the children's education.

However, it must be said that the community retained its pristine state by virtue of its isolation from a rapidly changing world. People were content with their standards of health, education and material well-being, but perhaps only because they were unaware that greatly improved standards of health and education services and of material goods were available elsewhere.

Araqi, then, was a good example of a developed traditional rural community. Its people were interdependent and thereby provided each other with most of the goods and services that they required. Economically, if one includes the Bedouin in the environs of the village, Araqi was almost independent of the outside world, the inhabitants having only a small need to sell goods with which to purchase the simple range of commodities they required that were not locally available. Within the range of technologies available to them the people made full and efficient use of locally available resources, they lived in harmony with their environment and they exercised appropriate control over the equilibrium between taking from the system in order to provide for present needs, and giving to the system in order to provide for future years and future generations. The village system was in harmony and balance with itself. Because of this harmony and balance Araqi's rural community might, therefore, be described as having been fully developed. Also, the passage of time had proved that the community system was truly sustainable.

2.2 Oil-induced change

In practice, of course, the situation of our idealised village was never as simple as described above. No rural community is ever fully independent, and the expectations of its members are subject to change by exposure to different ways of life and standards of living in neighbouring communities. In the case of Araqi (and the other villages of northern Oman) the ability of the community to respond to changing circumstances and remain a community was being put to a very severe test. The new oil economies of the region, starting in Bahrain in the 1930s, gave employment to many men from northern Oman and this experience rapidly and profoundly expanded their material, educational and health aspirations.

In theory there was no reason why the flexible and responsive control mentioned earlier could not have been used to reset the dynamic balance between input and output in Oman's rural communities to yield a higher level of production, if the people so desired, thereby enabling individuals to take more from the system to satisfy higher aspirations whilst ensuring that a correspondingly greater input was returned in order to sustain the system for the benefit of future generations. Greater output could have been achieved by harder work, by more productive work, or by undertaking new types of work. In this way the rural communities could have undergone extensive modification and yet remained communities. However, this would have involved an evolutionary process of adjustment that took time. Any modification to a known system involves uncertainty and may be harmful long-term even if seemingly beneficial short-term. Therefore change needs to be slow so that the impact of change can be appropriately monitored and evaluated, and modifications made to the changing system where and as appropriate in order to ensure a new, beneficial and sustainable dynamic balance at a higher level of production.

It is characteristic, however, that where societies in the Arabian peninsula, including those in rural Oman, have moved from an economic environment without oil to one with oil, the change has been very rapid. There has been no time to adjust to new circumstances. Change has been thrust upon the rural communities from without, by forces beyond their control, with consequences that have been in some ways harmful to the communities and to the management of the resources upon which they previously depended. This point is illustrated, below, by the results of a study undertaken by the Durham University team in the early-mid-1970s of the village and pastoral communities in its cross-sectional study area of northern Oman. At that time the pace of change was rapid and it was possible to see how change was dislocating the rural communities. The dislocation was visible because it was still possible to see the remnants of the traditional structures and functions of the communities as they had existed pre-oil, as the above idealised picture of Araqi indicates.

The descriptions of the changing social and economic life of the rural communities are given at some length. Quite apart from the insights they give about rural communities falling into disarray, they are a fascinating record of rural activities which, in some cases, although common in the mid-1970s, now no longer take place thereby illustrating the pace and extent of change with which the rural populations have had to cope. Perhaps the most remarkable feature of rural Oman today is the way most individuals have absorbed, with apparent acceptance, a rate of change which has transformed their lives in only twenty or thirty years.

2.2.1 Human resources

An estimated 26,000 people lived within the study area, and almost 16,000 of them were individually enumerated. Easily the largest settlement was Ibri (5,800) in the Dhahira with al-Khabura (about 3,400) on the Batina coast in second position. The average settlement, with 590 people, was much smaller whilst 80% of the settlements were below average size. There were a mean of 6.4 people per household, and in 16 of the 19 groupings (including village, Bedouin and *shawawi* communities) the mean lay between five and seven which implies that a very reasonable approximation of the total population size could have been attained by counting households (Birks, 1984a).

The average age of the population was only 21.2 years, but even more significant for population growth rates since that time, 46% of the population were under 15 years. Indeed, the crude birth rate, at 50·7 per 1,000 persons, was already very high. It was estimated that the rate of natural population increase was rising to 3% annually, and this in spite of a very high infant mortality rate of 196. During the 1970s and 1980s the enormous improvements in health facilities, resulting primarily from the rapidly expanding government health services, cut the infant mortality rate sharply which will have brought the rate of natural population increase well above the 3% level. This of course has tremendous implications for the provision of health, education and other services, for rural employment (or rural-urban drift) and for the demand for scarce rural resources, notably water.

Some employment in the mid-1970s was simple and full-time, but some was also dual, periodic, seasonal, informal or part-time. There were a total of 4,672 economically active people amongst the enumerated population, of which most were male. A quarter of them had only occasional employment. Farmers, pastoralists and *bidars* totalled only about 13% of the labour force, in this a strongly rural region.

Paradoxically, there was both a labour shortage and very considerable under-employment. Already the labour shortage (a requirement for skilled labour in the growing number of garages, for example) was being made good not by bringing more local people fully into the economy but by the importation of Asian labour. By the 1980s, this Asian labour had entered almost every aspect of village economic activity, from tailors and hairdressers to mechanics and farm labour, though not fishing (Dutton, 1987a). But at the same time as Asian in-migration was increasing, the out-migration of the locally under-employed was already very strong, and remains so today. In 1974 no fewer than 74% of the adult males aged 14-39 years were away working; the great majority of the most active age group (Birks, 1984b). About half the out-migrant workers were in the military, learning skills which were of no great value to their village economies but sending home remittance money. The next major employer of out-migrants was the petroleum industry (12%) and a further 10% had casual labour. In terms of destination, Abu Dhabi, came first (easily reached from the Dhahira, and the home of the Abu Dhabi Defence Force, a major employer), followed by Oman's capital region, then Fahud (oil), other places in Oman, Kuwait and Bahrain. Although the absolute majority of migrants came from the larger plains villages, the mountain villages gave a higher proportion of their total male population - indicative of a movement from the mountains to the greater work opportunities and amenities of the plains. Labour migration was associated with a transformation of the pastoralist and agricultural economies leading in some areas to a big fall in the number of livestock. It was also associated with a reduction in the level of skills in agriculture and with a partial breakdown of the communal *falaj* system as people used their remittance money to sink their own wells. Crafts also suffered. But to some extent remittance money benefited the local economy through investment in a range of small enterprises, from farming and fisheries to new shops, restaurants and garages.

Up to 1970 education in rural Oman scarcely existed, other than that provided by village Koranic schools, but thereafter primary and secondary education services expanded very rapidly. In the study area the first school opened, in al-Khabura, in the academic year 1972/3. The first intakes included many older boys (and girls) who had not had previous opportunities for education. But although excellent in principle, the new education system had several disadvantages (at least in the short and medium term) for rural community structure. First, all the teachers were expatriate (mostly Egyptian) and were unable to relate their teaching to local social, environmental or economic issues. Second, the school children aspired to regular employment in the armed forces or the national government because they offered an easy life and a regular salary, and so were lost to the rural economy. Third, the children spent much less time with their parents during the working day and did not imbibe their inherited understanding of or concern for the proper management of rural resources. It is a sad comment that one of the Khabura Project's best local workers, who became an excellent paravet, stayed with us because he repeatedly failed his local school exams and therefore could not move to the town to finish his education and find a job. More profoundly, an ever-diminishing lack of awareness of the critical value of conservation, coupled with other forces acting within the changing economy, made people more careless about water usage and the felling of trees.

2.2.2 Livestock

Livestock in northern Oman were (and are) owned by *shawawi*, Bedouin and villagers. Along with fishing and the cultivation of the date palm they were a main stay of the traditional economy.

The *shawawi* were defined as the non-village, non-farming population of Northern Oman, excluding the Bedouin inhabitants of the interior plains. By tradition they were mobile and owners of goats and sheep but by the mid-1970s many *shawawi* families owned fewer small-stock than the average number of goats and sheep owned by village farmers, and many of the *shawawi* households were little more mobile than those of many villages. It was found that *shawawi* life had changed least in the mountains. The remoteness of these *shawawi*, and the absence of neighbouring large villages which elsewhere acted as a focus for change, militated against their rapid social evolution (Birks, 1982). However, increasingly towards the end of the 1970s the growing number of graded roads in the mountains, and of vehicles using them, brought even the mountain *shawawi* into close contact with the large village settlements on the plains. Then the severe three-year drought from 1977/8 increased pressure on mountain *shawawi* and villagers to move down to the coast where water for agriculture and domestic usage, and other modern amenities, were available.

There were more goats per household in the mountains than on the plains; 57 on average in contrast to 27 and 32 on the Dhahira and seaward pediments, and of the total goats enumerated 7,000 of the 12,000 were in the mountains. For a combination of these two reasons the mountains were the most important area of pastoralism in this part of northern Oman. Some 55 households owned 50-300 goats, which indicates that pastoralism was making a significant, and not just a marginal, contribution to the economies of over 40% of the mountain *shawawi* population.

In the mid-1970s it was noted that for these mountain *shawawi*: 'considering the large numbers of small stock which are kept in the mountains, a surprisingly small amount of effort is put into their upkeep' (Birks, 1982, p. 5.9-12). Most of the time the herds were left to fend for themselves, and a significant number became lost, often ending in their deaths by predators, falls and, in extremis due to drought, eating the plant *Acridocarpus orientalis*. In addition, it was estimated that up to 60% of the kids might be lost each year due to ill health, poor nutrition and other management factors.

In the Dhahira the *shawawi* formed only 8% of the population and each *shawawi* household owned on average only half as many goats as their counterparts in the mountains. Therefore livestock assumed a less important role in the Dhahira, and indeed in the mid-1970s 62% of *fariqs* (*shawawi* households) owned fewer than 25 head of goats. Moreover, in the year prior to the survey 40% of the *fariqs* had not sold an animal, and some had sold no animals in the previous five years. Goats were being kept mainly for the *shawawi* themselves to eat, and their income earning potential was limited and on the decline. There was strong evidence that as labour migration to the Gulf had increased in the previous decades so the flock sizes were being reduced. In the case of some Bani Qitab groups, for example, flock sizes were estimated at 'only a fraction of the numbers' (Birks, 1982, p. 5.9-33) owned 20 years previously, as a result of the large number of men employed in the Gulf in the interim. Gulf pay packets had become so attractive that many men were moving away from their *fariqs* so that their pastoral system of husbandry had begun to break down. Wells were falling into disrepair, animals that strayed were not recovered, and the men were not available to organise *fariq* moves. As it became more difficult to look after the animals it seemed easier, when meat prices rose during this period, to sell the animals for slaughter and begin the transition to a more sedentary lifestyle. Traditional communal structures as well as systems of animal husbandry were falling into decay. The same picture of pastoral decay related to absentee migrant labourers was found also on the gravel plains of the Batina. The *fariqs* located near the Batina agricultural zone had the highest rate of absenteeism, a lack of men which was clearly responsible for the eroded pastoral economy of the *fariqs* in this area. No Batina *fariq* head interviewed owned more than 30 goats, none was primarily a pastoralist and all were reliant upon some other source of income.

On the plains rather more attention was still being given to the animals than in the mountains, and although in part this was merely to keep them away from farmland the goats were nevertheless given some supplementary feed in the form of alfalfa and dates. But overall we are dealing with a minimal management system; whilst overt costs of production were very small, productivity was comparably low, and attempts to improve productivity were negligible. There were also high hidden costs. There was no possibility of improving productivity through selection or breeding, there was no effective protection against disease, a high proportion of kids died, and kids and adults were lost to predators or injury. Very importantly, whilst animals were keeping themselves alive, they were yielding very limited returns at the expense of rangeland degradation.

Oman's pastoralists, however, were not ignorant of the toll that browsing and grazing herds were making on the range habitats. Pastoralists and villagers developed, over years of experience, their own conservation practices. The most important of these is that trees were never cut for fodder or firewood. Fodder was instead collected by beating the tree with a large pole so that fruits and leaves fell and were collected in a blanket placed on the ground. Firewood was only obtained from dead trees or by lopping branches of living trees.

These practices mean that trees remain fairly abundant in many areas of Oman. However, by the mid-1970s there was evidence that the rigour with which these traditional conservation practices were upheld was diminishing, probably because the pastoralists, most of whom also obtained elements of their total family income from migrant labour, were not so dependent upon their livestock as before and were therefore less concerned about the long-term viability of the rangeland. It was noted, on the Batina for example, that many trees were being felled to make room for an expanding area of farms although many of the new farms were of doubtful economic viability. Detrimental lopping, cutting and felling of trees was also observed in the *Prosopis cineraria* (*ghaf*) woodlands of the Sharqiyah (Brown, 1988). It seemed that the new generation of pastoralists and other dwellers in rural areas of Oman were being more careless about protecting the rangeland than their fathers because they were growing up in an oil age when pastoralist livestock and the browse on which they depended had become of only secondary importance to the family economy.

A detailed study was also undertaken of village livestock in al-Khabura and in the coastal villages for ten kilometres either side of it, and the findings compared with the village of inland Hugayra on the Batina plain, and the small settlement of Harmali high up in the valley of the Wadi Hawasina (Dutton, 1982f; Figures 1 & 2). Many cattle, goats, sheep, and chickens were kept within the village households. Cattle were found within one-third of the house compounds, whilst three-quarters of the houses owned sheep or goats, or both. On average each house had 1.4 goats, 0.9 sheep and 0.5 cattle, but considering only the people who kept the animals the average per house was 2.3, 2.6 and 1.4 respectively. Although there were only about four sheep to every six goats they were kept in rather larger numbers by fewer households. In inland Hugayra there were more goats (7.1 for those houses with goats) and sheep. Harmali had above average goats but almost no sheep, reflecting the dominant position of goats in the mountains.

The problem in the villages was not so much of a livestock system in decay, as a livestock system almost entirely geared to supplying the households with only occasional supplies of meat (particularly to mark the two main religious festivals of the year) and therefore a wholly inadequate basis upon which to build a productive livestock industry. Goat and sheep breeding was haphazard. For 125 of the 442 pregnant sheep and goats in the survey, information was obtained about where the animal had been served. In 113 cases, some from each of the villages, the animal was simply served as it wondered freely amongst the houses; no attempt being made to select the male nor to control the breeding frequency. The remaining 12 animals were left with the *shawawi* for one or more months to graze with the their flocks and be served, in the belief, or hope, that the *shawawi* males were of better quality. However, the stock holders on the plains, and those questioned in Harmali and elsewhere in the Wadi al-Hawasina, where prolonged drought had reduced the grazing to a minimum, were complaining that the females were too hungry to become pregnant when served, or else were not able to carry the foetus full term. This reveals the minimal-management system in its worst light; because of drought there is very limited grazing or browse, therefore a high proportion of the browse that is available is eaten by the goats causing degradation of the range. However the quantity eaten is still so small that it merely keeps the goats alive without producing any meat and even prevents reproduction.

In six of the villages of the study area questions were asked about illness and deaths of livestock. In the previous 12 months only a few cattle had died, some of these from accidents, and a large number of chickens. However, much more numerous than cattle deaths, and economically much more important than chicken deaths, were the mortality rates of both sheep and goats. During the year 1974/5 some of the villages were much more badly affected than others, and in each village goats suffered more than sheep. Deaths expressed as a percentage of remaining livestock numbers varied from 4%, 5% and 27% (Qusabiyat al-Hawasina, inland Hugayra and al-Rudayda), to 57% (Khor Rasal), 70% (Muhaydhib) and a devastating 158% in Qusabiyat az-Za'ab. With the threat of these high levels of mortality hanging over the stock owners there was very little incentive to invest in building up larger or improved herds. But although people frequently noted that successive animals died within a few days of each other, and it was generally understood that disease could be transferred from one animal to another, no attempt was made to isolate diseased animals or to take healthy animals away from an area of infection.

The seriousness of the problem may be gauged by comparing the value of the animals which died with local wage levels in the mid-1970s. Three goats, at the prices pertaining in the *suq as-Sabi'* of December 1974, would be valued at RO105[1], which was equivalent to at least two months salary of most of the people employed locally in agriculture. So although people were not aware that they themselves could help reduce levels of disease and mortality, they were nevertheless very concerned by the wastage of livestock and asked repeatedly whether veterinary services could be improved. But any kind of service to improve livestock husbandry and productivity was made difficult by the fact that the animals were scattered between so many village homes and allowed free range. Access by veterinary or other specialists would have been almost impossible, particularly as the men were increasingly working away from home and so the specialist would have had to be received by the women - very difficult in Oman's traditional rural society.

Livestock in the villages had a better diet than those owned by the *shawawi*, or at least one which was much more independent of cycles of rainfall and

---

[1] Rials Omani (RO): RO1 approx. £1.5

17

drought. Some of the cows, for example, were notably well cared for, and both goats and sheep were typically fed some alfalfa and local dates on a regular, often daily, basis. In addition, the goats were fed weed grasses (cleaned from the alfalfa fields), leaves from trees, fish heads and entrails, stale bread, scraps of cooked rice and other odds and ends. Unfortunately, however, the goats, while wondering through the villages, picked up other scraps including orange peel, old cardboard and other materials which had limited nutritional value and sometimes caused alimentary canal problems.

Indicative of the subsistence economy approach to livestock husbandry in the villages was the *binus* system used by people who wished to acquire an animal in full ownership but had not the money to buy it in one go. They would take, for example, a female calf, rear it, have it served by a local bull and see the cow through its pregnancy. The calf then belonged to them. The cow, typically, was retained until it had finished suckling the calf, then sold and the money for it divided 50-50 between the owner of the original calf and the owner of the new calf. In the case of sheep and goats the system was very similar, though a rather higher proportion (than the cattle) were male, normally fattened over the course of a year with the intention of selling before the *Id al-Fitr* or the *Id al-Adhha* and then dividing the money. The principal benefit of the system to the acquirer of the calf (or kid/lamb) was the subsistence advantage that he could rear the animal without any major single financial outlay; it was easier to pay a little money each day for the food (even though the total cost may have been more) than to buy the animal outright.

It has been stated above that the village livestock system was not in decay, but this is not entirely true because during the 1970s the animal by-products, particularly wool/hair and skins, were being less and less utilised. In 1974/5 in the coastal villages, of 223 people who said they had wool, 220 threw it away, none used it, and only three reported making any sales of wool, to the *shawawi*. People reported that there had been a demand for the wool in the past, by the *shawawi* women who span it. Even in inland Hugayra, a loose settlement of *shawawi* origins, only one person out of eight with wool used it for spinning. There were no sales, but a few women in the inland settlements still span when they wanted a rug (*mansul*), taking the required number of balls of yarn to the al-Khabura weaver (male) who made the *mansul* on his pit loom and charged RO3 for his labour of two days (Dutton, 1982f). But even the *shawawi* with *mansuls* in their houses also owned brightly coloured rugs bought by their husbands who were working in the Gulf. One woman said that although these rugs were more expensive (RO4-5), less durable, less warm and less waterproof than the *mansul*, they were prettier and less work, and quicker and simpler to obtain (Dutton, 1983a). Thus the weaver was getting less work, and at the same time his sons were remitting money back into the family from work in the Gulf. The inevitable consequence was that the weaver finally stopped work in 1976, and there was, thereafter, no point in the women spinning.

As for wool, so for skins. Of the 560 houses with skins in the 1974/5 survey, 86% threw them away, 10% used them and 4% sold some of them. Most of the goat skins used were made into milk churns, two needed per cow for each lactation, used on alternate days. Thus in inland Hugayra, where there were more cows per household than in any of the coastal villages and therefore more need for milk churns, 73% (16) of the sample used some of their skins for churns and only 18% threw them all away.

At the time of the survey there were still, in al-Khabura, two professional leather workers (*shammar*) but they were elderly men and complained that there was now very little work for them to do, though they still made some churns. But before the introduction of diesel water pumps in the gardens of the Batina during the 1960s, there was a steady market for animal skins with which the *shammar* made leather for the huge trip-buckets of the *zagiras* which hoisted water from the wells. Additionally, small quantities of leather (cattle and goat skins) were used to make sandals, and some sheep skins were tanned to make 'saddles' for donkeys. Furthermore, there had been a strong if fluctuating export market for hides but by the 1970s this trade had virtually ceased (Dutton, 1982f).

Leather making was in fact very skilled. The pods of a tree, *Acacia nilotica* (*qarata*), were the source of the tannin, and a mixture of salt, pods and dates was crushed, beaten together, spread over the flesh side of the skin, left for 7-8 days in the shade and finally scraped off together with the wool/hair which was loosened during the tanning process. The remarkable features of this technique were that it was virtually dry, that no lime or lime substitute was used to break up the fibre structure in order to allow ingress of the tannin and that the wool/hair was successfully loosened by fermentation-putrefaction without putrefying the skin. Indeed, by a subtle variation of the process, when a donkey saddle was needed, the wool was left firmly attached to the skin. The tanner also sewed together the *dellu* or churn to make a watertight, or milk tight, container. The churns were both strong and subtle, having to withstand a rigorous process of cleaning with salt and water after use.

Thus the fact that most of the skins were being thrown away meant not only the waste of a local resource and a reduced income for the livestock owner, but also the passing into history of an age-old but sophisticated craft skill. The result was the conversion of skilled rural craftsmen into unskilled (and unemployed) labourers because no one any longer had used for their inherited expertise.

In the case of two other animal by-products, milk and dung, it was not so much that their limited usage reflected a system in decay, more a subsistence economy mentality in which the exploitation of these by-products had never been fully developed.

The *shawawi* goats, mostly free-ranging, dropped their waste products as they travelled making their collection difficult or impossible except where the animals assembled by the *shawawi fariq* in the evening. Even then the dung was only infrequently collected and used. In the coastal villages dung collection was easier. Most of the village livestock were normally kept in the house or in an adjacent palm-frond pen. The sheep and goats were allowed some freedom to wander within the house and between the houses of the village but cattle were normally confined to a pen within the house compound and were usually tied to a stake by one leg as well so that all their dung and urine fell in one place. Dung from the house - both cattle and sheep-goat dung - was often collected and thrown onto a dung heap or midden outside together with sand and ashes. The dung then baked hard in the sun and no attempt was made to rot it down. In some cases dung built up in these middens for years and donkey dung added to the pile from a donkey tethered on the midden itself. Sometimes dung from the midden was used as required or sold. Sales were always on a casual basis - if a buyer happened to arrive - and only occurred once or twice a year at the most, for about RO1-6. In fact, in the coastal villages three-quarters of the people with dung threw it away, about 20% used some of it and only 5% had made sales during the previous year. So dung was a greatly under valued and under used resource.

Milk was at least a valued product. Only cows were kept specifically for milk, but in many instances a lot of care was lavished on them, almost like household pets. Careful questioning of the cow owners, and volumetric measurement of the containers they milked into, suggested that a few cows at the height of their lactation gave at least five litres a day but that 50% of the cows (24 of the 50 for which yield information was obtained) gave under one litre per day. Some of the milk was fed to the calf, and all the remainder was used in the household - in the study area no example was discovered of milk or milk products being sold. A little milk was drunk in tea but most of it was converted into butter (*dehen*) and buttermilk (*leben*). To do this the milk from one day was collected in a large bowl (perhaps kept under a larger up-turned bowl to prevent dust fouling it) and overnight a starter, obtained from previously treated milk, was added to it. Very early in the morning the milk was all transferred to a churn (made from a goat-sheep skin), suspended from the apex of a wood tripod and shaken for 1-2 hours until the butter separated from the buttermilk. The latter was drunk by all members of the household, and the butter was similarly used in the home. Very occasionally a cream/cottage or curd cheese was made. The cheeses were crude (the curd cheese was neither compressed nor matured, and retained its rubbery curd texture) but otherwise the processing of the milk was undertaken with very considerable skill and, in dusty and hot conditions, with an empirical understanding of hygiene. But it remained a subsistence activity, no one further developed or exploited their dairy skills with the intention of making sales.

Another highly developed local skill that depended on animals was the use of bulls for ploughing. However, bull ploughing, at least on the Batina coast, disappeared during the late 1970s when the bulls were made redundant by the extension centre tractors. But in 1975 six ploughmen were still active on the coastal gardens in the *wilaya* of al-Khabura, each using a single furrow plough pulled by two bulls. They either in an anti-clockwise movement ploughed the small circular basins round individual date palms, or else ploughed parts of open fields. Fields were always ploughed twice, at right angles, the second ploughing being undertaken after the seed were broadcast.

The ploughmen said then that they still had work for several reasons, in spite of the extension centre tractor: because people had to wait a long time for the tractor, because only the wealthy people were able to have use of the tractor, and because the tractor could not work in restricted areas round and between the date palms and other fruit trees. But the ploughmen even then did not seem to work full time. One reckoned that he worked on average every second day during the winter, mostly for the wealthier people ('the rest do not bother to plough round their date palms any more'), but in summer, although he could find work, he stopped to tend his date palms (Dutton, 1987d).

The yokes and the ploughs were made locally by a woodworker in Rudayda (workshop and house in his palm garden) and the metal tip of the plough was made by a blacksmith, working near the *suq* in al-Khabura. Cuttings from local trees were used to make the equipment; the plough from *Acacia nilotica* (*qarata*) which is very hard and durable, and the yoke from *Ziziphus spinachristi* (*sidra*) which is more flexible and resilient. By taking all prices and costs (including food for the bulls, which was the major cost) into consideration it was calculated that if the ploughman worked for 200 days a year he made a net income of RO21.5 per month. By the mid-1970s this was regarded as only a very small income, in comparison with salaried incomes to be earned by moving away from the village, but the ploughman was unable to increase his charges because he had suddenly found himself in competition with the very costly but almost fully subsidised extension centre tractor ploughing service, and this, by the end of the 1970s, was the factor which finally put the ploughmen out of business.

The above is an example of one part of the traditional system being overwhelmed by externally introduced change, change which gave no thought to the merits of the traditional system and, through subsidy, put it at a grossly unfair disadvantage. The argument here is not that tractors should not have been introduced, but that they should have been introduced in a way which allowed the different strengths of the tractor and the bull plough to demonstrate themselves. It is likely that the tractors would have found a role on the larger, open fields of the new farms whilst the bull ploughs might have remained more suited to the small and restricted patches of land on the old gardens. Both tractors and bulls were timed working in restricted areas. The bulls averaged 40 hours per hectare, and the tractors 8-11 hours per hectare. This is extremely slow for a tractor, emphasising how unsuited they are on small areas of land.

And if one also considers that they do more damage to the irrigation network and to neighbouring plots, and if one takes into consideration the time spent moving from plot to plot and farm to farm their utility was very low indeed.

It would have been worthwhile making a more detailed study comparing bull-ploughing with tractor ploughing with a view to encouraging the former in certain conditions. The advantages of bull-ploughing were many and varied in that the activity: used locally available skills and craftsmen; did not rely on imported machinery, fuel and spare parts; was extremely flexible in operation (suited to work between palm trees and in small areas of ground); did not require costly mechanics and back-up service; spent a greater proportion of its working time on the job; and was equally suited to large and small settlements and to coastal, mountain and interior communities. But there was also, potentially, room for a third approach to the problem of ploughing, one which lay between the extremes of the bull and the large tractors. It was likely that a pedestrian controlled rotavator would have been small enough to work more efficiently in the smaller gardens. Being smaller and cheaper than the tractors more rotavators could have been provided thereby reducing the time lost in travelling from garden to garden. But it was clear that such an approach to ploughing would need to be introduced by a process of demonstration and the encouragement of private, direct involvement so that the control of the system remained squarely in local hands. The practice of providing large tractors at very subsidised rates meant that they not only acted in very unfair competition with the bull-ploughmen but also inhibited the introduction of other tractors or rotavators by private enterprise. And in the process a number of interrelated and mutually dependent skills associated with bull-ploughing, one part of the fabric of local community structure, was thoughtlessly cast aside.

2.2.3 Agriculture; the date palm
By the mid-1970s change was overtly in hand in the gardens of the central Batina. The agricultural surveys divided the gardens, primarily, into 'old' and 'new' which roughly equated with pre-oil and oil times. The old gardens were small, near the sea where were located the villages and where the water table was highest, and were dominated by date palms. The new gardens were inland of the old gardens, alongside and inland of the new main road. They were also larger and with practically no date palms. The main crop was alfalfa, which remained clearly profitable, and the owners of the new gardens were also experimenting with a wider range of vegetable and other crops, including bananas. The lack of interest in date palms represented a profound change in people's attitudes to this crop which had for ages been of central importance economically, nutritionally and even culturally to the local community. In the old gardens no less than 83% of the land under crop was planted with date palms, but in the new gardens it was under 4%. What had happened to cause this profound change, and what were the consequences for the community? (Dutton, 1982d).

The answer to the question about cause lay in the bottom line; the Batina date gardens were no longer profitable or otherwise beneficial to their owners. But this is a statement which requires some explanation. One of the problems, in an age when other, more remunerative and less arduous work opportunities were becoming available, was that date palms were very labour demanding. There is a long list of jobs that have to be performed on a regular basis, usually annually, to maintain the health and productivity of the palm. Each operation is done manually, the worker climbing the tree, up into the fronds and amongst the thorns in some cases, in order to complete the task. Each job demands experience, skills and care, and carries a certain risk of falling or of being pierced by the thorns, which are long, strong and sharp-pointed. The first task, cutting off the dead fronds, is the least skilled work and may be done on a casual basis when people need fronds for making *barusti* sheets or when they need to cut through the old fronds in order to climb up to the top of the palm tree. The fronds are cut with a serrated sickle leaving a butt (*karab*) perhaps 25cm long still attached to the trunk. The second task is the *khalab* in which a specialist, the *khallab*, removes the butts. He climbs, with the protection of a sling tied round the trunk of the palm, using the stumps of the previously cut butts as footholds, up to the level where frond butts remain. He trims back the butts using a special sharply-curved and pointed sickle, and he also cuts away the associated date palm fibre.

Completing the *khalab* allows the palm to be climbed more easily when the remaining tasks have to be performed but its main value lies in removing a niche in which debris may lodge and from which palm-boring larvae of rhinoceros beetles easily penetrate the palm trunk. Although some farmers undertake their own *khalab*, more typically it is the work of the specialist *khallab*. In Araqi in 1974/5, for example, four brothers, the Awlad Humayd, seemed to have a near monopoly and in every village visited in the Dhahira *khallab* were responsible for most of this work. Without exception they were all said to work at the rate of 200 butts/riyal. At that rate, and assuming that the average palm has 15 butts, a 100-tree palm garden would have cost RO7.500 to trim. As the *khallab*, according to their own statements, could trim 30-40 trees per day, they were earning about RO2.500 per day. On the Batina the *khallab* were employed rather less frequently - the owner sometimes doing the work himself, or the work not being done because of the relative neglect of some of the older gardens. The average of eight informants gave the price of RO7.500 per 100 trees, which was equal to the rate in the Dhahira.

The third, and perhaps most critical and skilled task of all, is pollination. Date palms are either male or female. If wild they would be wind pollinated, with roughly equal numbers of male and female palms to ensure effective pollination. However, if the pollen is transferred by man from the male flowers of one tree to the female flowers of another, pollen wastage is greatly reduced. In this way a garden with a hundred palms can have 97 or 98 females, and only two or three males, and thus the yield per hectare is greatly increased. But good pollination requires both skill and experience, and difficult and dangerous work. The specialist first has to climb the male palm to cut down the maturing male spathes. He then has to climb each female palm, up through the fronds, to reach the new female inflorescences and place in them parts of the male flowers. The technique involves climbing each palm twice or even three times, for earlier and later inflorescences. In the Dhahira two types of pollen are used, one (*nebat gharif*) being required for the date variety, *fardh*. For some varieties fresh pollen is used, in other cases it is dried first. There are also other variations on the general technique for different varieties in different circumstances which are only learned and understood by long experience. In many villages of the Dhahira pollination is normally done by the *bidar*, the same man who irrigates. In fact, being fully skilled in pollination is one of the main qualifications a *bidar* is expected to have. Estimates of cost varied: 100 baisas/tree/climbing, or 250b for a small tree, or up to 500b for a large tree. On the Batina pollination specialists were also used, though more often than in the Dhahira the task was performed by the owner. Eight cost estimates averaged RO5.250 for 100 trees - much less than in the Dhahira but probably accurately reflecting the fact that the Batina palms have on average many fewer hands of dates per palm than in the Dhahira.

The next task, once the dates have started to grow, is tying down the date hands in order to make subsequent harvesting easier with less likelihood of dates falling and being lost. There are three ways in which the hands may be tied depending on the variety and on whether the dates are to be picked and eaten fresh, or to be dried for winter storage. The *tahdir* involves bending down the stem of the date hand and tying the hand to one of the lower fronds. The dates are much less likely to be blown away by high winds, and the whole hand can be cut and removed without fear of rubbing off the dates against the fronds. However, the method *tasgir* whereby the hand is brought very low on its stem and the base of the hand is tied to the frond near the butt, is utilised for the higher quality palms. Finally the *tahzim* involves bundling together the parts of the date hand just as the date is maturing. This prevents the mature dates falling and allows them to dry on the tree. On the central Batina *tasgir* was unknown because no higher value palm trees are grown there.

When the dates are mature, harvesting is the culminating task in the annual cycle of seasonal date palm work. The dates may be harvested in one of two ways: *kharaf* means picking individual dates to be eaten fresh; *gidad* means cutting the whole hand and is normally associated with preparing dried dates for winter storage. A few estimates of the cost of the *gidad* were obtained, averaging 250 baisa/tree in the Dhahira, and RO6-10 per one hundred trees on the coast.

Finally, at the end of the date harvest and again in the spring the land beneath the palms was ploughed. The spring ploughing, in the *falaj* gardens, prepared the land for summer forage crops, while the post-harvest ploughing cleaned the land after the forage had been cut. The work was undertaken by specialist ploughmen who trained their bulls to work in very restricted spaces and close up to palms and other trees. On the Batina, only the small round irrigation basins encircling the palms were ploughed.

In addition to the annual cycle of tasks, was the continual schedule of irrigation, whether from the *falajes* in the Dhahira or from hand-dug wells on the coast. In the Dhahira, irrigation was the routine work of the *bidar* whereas in the central Batina it was usually the responsibility of the owner of the garden. In the interior of Oman including the Dhahira the *falaj* system of irrigation had remained essentially unchanged, but on the Batina, during the 1950s and 1960s most farmers replaced their bull and trip-bucket system of raising water (the *zagira*) by diesel pumps. Other people with date garden responsibilities in the Dhahira included the *wakil* and the *haris*. A *wakil* or agent was appointed by some members of the Ubriin tribe living in Hamra (near Nizwa) to take general responsibility for their land in Araqi. The *haris* (guard) was typically a poor man who guarded the palms against theft and animals and was permitted in summer to grow forage crops beneath the palms. But in the settlements of Ibri and Sulayf the *haris* had full responsibility for the date palm work except for irrigation. On the Batina there was no one strictly equivalent to the *bidar* or *haris*, but in a few cases an *'ammar* lived in and looked after the garden of someone who was working in the Gulf.

A large input of skilled labour was, therefore, essential to maintain a date garden at high and sustainable production levels. Information collected on the costs of each of the cycle of tasks, including irrigation, ploughing, *khalab, nebat, tasgir* and *gidad* showed that in 1975 a Batina garden containing 100 mature palms cost RO97 annually. At the same time data obtained from 21 farmers suggested that a 'high average' yield on the Batina was only 3.6kg per palm with a market value of RO108. The farmer could add to this income some RO24 by selling *barusti* sheets giving a total income of RO132 per 100 palms, and a negligible net income of RO35; and this was for a farm with 'high average' yields of dates. The majority of farms were losing money. An important reason was that the variety *umselli* which dominated on the coast and was tolerant of the relatively poor coastal conditions for growing dates, gave only a lowish date yield of distinctly poor quality. As Omanis became more wealthy resulting from regional oil wealth they were diversifying away from dates in the foods they bought. In the process the poor quality Batina dates were largely rejected, except as animal food. The position in the Dhahira for a garden containing one hundred mature trees was rather more optimistic. The cost of irrigation, ploughing and all contract work was higher at RO199, but the income from the dates, assuming that they were all sold on the tree (*tana*) was RO460, leaving a net income of RO260 annually. The yields were higher as were both the quality and the price.

In the Dhahira a combination of higher demand, the inertia created by a combination of *falajes* and tree crop, and substantial government subsidy has, in more recent years, maintained the palm growing system largely unchanged. However, on the Batina, near al-Khabura, the poor quality and price of dates from the 'old' gardens has led to the total or partial neglect of many of them. The role of the date palm, which used to be central to the functioning of the rural community, was being marginalised. Skills long associated with good palm husbandry were disappearing. Worse still, the problems were being greatly exacerbated by the ever-diminishing role being played by other parts of the date palm.

It is an important fact that the palm is by no means merely a producer of dates. The trunk, the fronds and the leaflets all have a variety of uses, providing a resource for building construction and local crafts, as well as fuel for cooking. They therefore, as is indicated below, provided some additional money to the farmer and also an income to the many people skilled in working the raw materials into their final products.

One of the main uses of the palm fronds was in the manufacture of *du'un*, or sheets of *barusti*. The fronds were all cut to the same length, partially trimmed of their leaflets, steeped in water to make them pliable, laid side by side and knotted together using local rope to make a 'modular' building sheet measuring about 5m by 2.5m. Such sheets were used to construct houses and other buildings in most parts of northern Oman, but particularly on the whole length of the Batina coast where on the shoreline, being sandy, alternative building materials were in short supply. A framework for the house was constructed from local timber, knotted together with local rope, and the sheets of *barusti* were tied on to form the walls and the pitched roof. The sheets had the quality of net curtains in that you could see out, between the fronds, but people outside could not see in. Only the cooling sea breeze could enter, acting as a simple air-conditioning system. A team of five men, and a boy to pass the fronds to them, is needed to make a sheet, and the team may complete four or five in a long morning's labour. Thousands were required each year so their manufacture was a significant occupation or employment. A man owning 140 palms, the average for the Batina survey area, could make 11 *du'un*. In 1974/5 they sold for RO3 per *da'an*. By 1978 the price had risen to RO5 but from then into the 1980s it remained stable reflecting the fact that demand had dropped completely because all new building work relied on the by then ubiquitous concrete block.

The local rope used to knot the *du'un* together was made from the sheets of fibre that the palms produce at the base of each frond. The palm fibre is highly resistant to decay and the individual fibres have considerable tensile strength. They made strong, coarse rope, and the sheets of fibre had also more mundane uses such as packing in donkey and camel saddles and to line earthen gateways in irrigated fields.

26

To make rope the sheets of fibre were washed, soaked in water, and beaten to soften and separate the individual strands. The strands, still wet, were then twisted between the hands into short lengths and the lengths similarly twisted into a two-ply rope. A steady day's work yielded about 50m of rope worth, in 1975, 150 baisas.

Rope locally had many uses but perhaps more was used to make the *du'un* and tie them onto the house frames than any other purpose. The same rope, or perhaps a thicker three-ply version of it, was used in the construction of the palm-frond *shasha* (see below), the characteristic small fishing vessel of the Batina coast. A much thicker rope of the same material was the basis of the sling which supported the date palm specialist when at work. Also, local rope of course had many general purpose uses for farmers, fishermen and others. However, the trend during the 1970s was to use other, nylon rope for general purposes, and during the latter 1970s it became standard practice to use nylon rope in *shasha* manufacture. As concrete blocks replaced *du'un* in house construction so the main use for local rope disappeared. The rapid phasing out of local rope during the 1970s, after centuries of usage, not only caused a skill to die and replaced a local resource with imported nylon, but restricted the opportunity for women, for the elderly and for the disabled to have gainful employment and to maintain their independence within their village societies.

Palm fronds were also used to make *shashas*, the 'float-boat' characteristic of Oman's Batina coast. It is said that in the post-war period the beaches were covered in *shashas*, but although they were fewer in number by the middle of the 1970s, those that remained were still in regular use (Donaldson, 1978a). Typically the fishermen made the *shashas* themselves. They bought 140 or more palm fronds, soaked them in the wet sand, trimmed them and bound them, with date palm rope, onto a simple frame also made from parts of date palm fronds. The boat was completed with a deck of the same material, and beneath the deck the hull was stuffed with palm frond butts. There was no attempt to make the craft watertight; given buoyancy by the butts, the vessel simply floated like a boat-shaped raft. The boats were rowed or else driven before the wind by a crude triangular sail. Usually a fisherman owned two (or more) *shashas*. Because the butts slowly absorbed water they gradually lost buoyancy, so the second *shasha* was used while the first was left on the beach to dry out. The first change during the 1970s was the replacement of the palm frond butts with broken sheets of expanded polystyrene, which retained their buoyancy because they were non-absorbent. The boats were lighter to launch and a fisherman needed only to own one of them. At roughly the same time local rope was replaced by nylon rope for binding together the fronds that formed the hull and for the simple rigging. But paradoxically as fishermen became more familiar with outboard engines during the 1970s the traditional *shasha*, temporarily, was given a new lease of life. The inherent plasticity of *shasha* design coupled with the inventive skill of the fishermen-craftsmen who made them, led to the creation of *shashas* which were larger and longer and included a specially shaped hole near the stern through which the propeller shaft was lowered with the engine mounted in the rear of the craft - a kind of inboard-outboard engine. This skilful remodelling of the *shasha,* together with the fact that the aluminium boat imported to replace it proved to be unstable, extended the life of the craft by some years. However, fibreglass boats in more recent years have largely ended its days. Thus the *shasha*, whose skilled manufacture had been an important user of palm fronds, rope and butts, has disappeared, and its disappearance has further reduced the value of the date palm, caused the loss of an indigenous skill, and made the local community less self-reliant.

The date palm provided the raw materials for most of the wide variety of baskets and mats that were found in Oman. Almost all were made according to the same basic method. A belt was plaited from date palm leaflets (*khus*) cut from the frond either green or dried. If green they were allowed to dry for several days but then moistened in water to soften them. The moist, pliable *khus* were split into 1cm wide lengths, and these lengths, two to four thick, were plaited into a braid 2-3cm to 6-7cm wide depending on the type and size of the article being made. It was known what length of braid was required for each item, and they were plaited accordingly. In making the final product the braid was slowly wound round on itself so that successive loops met edge to edge. The edges of the braid were bound together by a string, made from the date palm, being threaded through the plaits using a needle (*miselle*) made from a sliver of date frond. Handles, for those items equipped with handles, were made from date palm fibre rope (Dutton, 1982b).

The wide range of products made in the above manner include the following. The *khasafa* are the standard sacks for storing and transporting dates in Oman. They were, and remain, one of the major uses for palm braid and provided significant employment, particularly for the elderly and the disabled. The *thug* was the double donkey pannier usually used for carrying either earth or animal dung, but today rarely seen. Similarly, the *qafir*, a general purpose farm basket with two palm rope handles for moving small amounts of earth etc. by hand, is today rarely used. The *simmat khabat* is a very large, handled mat that was commonly used for collecting the leaves from the *Acacia tortilis* (*semra*) trees when they were beaten from the trees for feeding to livestock. Other mats, including prayer mats, were made using the identical technique. However, of all the above baskets and mats only the *khasafa* (known on the Batina as the *gharab*) remained in common use throughout the 1970s.

The date palm is also a source of timber, albeit of a very coarse-fibred type. If a mature date palm was blown down in high wind or had to be cut down for any reason, the farmer could sell it to the local palm tree carpenter, the *gidhdha'*, who made roof beams and other locally used products from it. In the village of Araqi, for example, in the mid-1970s, one family of three brothers was particularly renowned for its skill in cutting up date palms. Evidently the *fardh* was the preferred palm variety because it was the strongest, though some other varieties were also acceptable. Using a special set of tools the tree was cut into lengths and then the lengths split into beams of triangular cross-section which were carefully smoothed with an adze. It was observed that two medium-sized palms yielded 20 beams (or *gidha'*) worth a total of RO28. But the tree cost RO8 and the work took five days, so the *gidhdha'* gained a net income of RO4 per day. However, during the 1970s there was a gradual shift from traditional to non-traditional materials for house building, and just as the concrete block replaced the mud brick, so the imported lengths of 2" by 4" Indian hardwood replaced the local palm timbers as roofing beams. Another skill, another local raw material, and another example of mutual self-reliance disappeared from village community life.

From the above description it is clear that the date palm, over a very long period of time, played a dominant role in the economy and livelihood of rural Oman. It provided food (dates) for both people and livestock, and it provided fronds, frond butts, fibre and trunks used to make a very wide range of products as central to the rural economy as housing, fishing boats and a variety of mats and baskets. Associated with the date production and the craft manufacture, the palm tree has determined the development and the inheritance of a large number of intricate skills used by farmers and professional craftsmen alike. The palms therefore have provided employment and remuneration for many people in rural Oman. In addition, through date exports, the palm has been a major earner of foreign currency. But the oil era has altered everything. Dates are not such an important part of the diet now that other foodstuffs can be imported readily. Exports were greatly reduced even before the oil era began, including the exports of boiled *mebselli* dates. Many of the products made from date palm fronds and trunks were either no longer used, or were in greatly reduced demand. Centuries old craft skills were no longer valued now that equivalent products could be imported cheaply.

The cash income from most date gardens in the 1970s was small, and even, on the Batina, non-existent. Therefore many date gardens, particularly those on the coast, were semi-neglected. But there is a great reluctance to root out date palms, even when they are neglected because they lose money, and as date palms occupied 70-80% of the cropped land and, on the Batina at least, tied up scarce resources of land, water and labour on a financially worthless crop, it made it much more difficult to develop other crop systems and their associated skills.

2.2.4 Agriculture: wheat and alfalfa
Up to the mid-1970s wheat was one of the two major field crops grown in the Dhahira, the other being alfalfa (Dutton, 1982e). However, after the mid-1970s the decline in wheat cultivation was very rapid so that by the years 1979/80 and 1980/1 only very small areas of wheat were being grown. This collapse in the cultivation of wheat involved a profound change in the cycle of agricultural

29

activity in the region, as the description below will make clear. The collapse began between the harvests of 1974 and 1975. For example, in the village of Araqi in 1973/4 of the 74ha of peripheral field crop gardens there were 25ha of wheat and 27ha of fallow, roughly half and half wheat and fallow in conformity with the two year wheat-fallow rotation. However, in 1974/5 there were only 21ha of wheat and nearly 32ha of fallow, marking the beginning of the end of wheat cultivation in the region. Let us take 1973/4 as a standard year. In that year the 25ha of wheat were divided between 72 gardens, a mean of only 0.35ha in each. Only five fields were bigger than one hectare. Because the fields were so small and had been formed, many years previously, each with its surround of *falaj* channels and mud brick walls, at a time when all operations of cultivation were manual or used bull power, access for even small machinery was almost impossible.

Still, therefore, in the mid-1970s the cultivation of wheat remained very labour intensive. The land was ploughed twice, at right angles, using a single-bull plough at a rate timed at about 40 hours/hectare. After ploughing, the wheat fields were divided into small irrigation basins (*gelbas*) surrounded by ridges and supplied with water along roughly shaped extensions of the *falaj*. The ridges were made by two-man teams using a bladed tool called a *graz*; one man dug the blade into the earth and the other man, heaving on a rope attached to the blade, pulled a blade-full of earth onto the ridge. This laborious effort was timed at 11-12 *graz* days per hectare. After the irrigation basins had been prepared, the wheat, mostly taken from the harvest of the previous year, was hand broadcast over the entire area. The land was then irrigated by *falaj*, on a cycle varying between 12 and 16 days or so for a period of four and a half months. The application rate was measured in Araqi, a village with two large *falajes*, at a mean of 22cm per cycle, so the growing crop had to cope with conditions varying between flood and drought within each cycle.

While the crop was maturing little work was done on it until the grain began to ripen, towards the end of February, when a guard (*haris*) was employed to scare away the birds. The small fields, the lack of alternative food, and the cover afforded by nearby date palms meant that birds could devour a high proportion of the crop, so the guards used a number of strategies to prevent this happening. The harvest, from about 10th March to 10th April, was big occasion. Up to eight men or more worked together in a field cutting the wheat close to the ground using serrated-edge sickles. The wheat was tied into bulky sheaves which were piled into small rounded stacks, head inwards as a protection against birds while the grain completed its drying over ten days or more. Meanwhile, a threshing floor was prepared from beaten clay mixed with chaff often within the shade of a large tree. As the first stage of the threshing operation the wheat heads were cut from the straw, using a sharp-edged stick, and spread over the threshing floor. Two lines of men standing opposite each other then rhythmically beat the chaff from the grain wielding whole palm fronds stripped of their leaflets. The resultant mix was then collected up, winnowed, and sieved to separate grain from dirt. The whole process was repeated, the wheat measured from a *mikyal* (a wooden container holding 1.6kg) into a basket and thence transferred into bags of 40 *mikyal* each.

Thus, the whole process of growing wheat was extremely labour intensive. At the time when our idealised rural community existed this was a virtue. There was plenty of under-employed labour in the villages, and the communal nature of the work, often with each farmer helping his friends, helped to strengthen social bonds fostered by mutual self-reliance. The most skilled contribution to the process was made by the ploughman who not only had to train and control his bull but also made his own plough from carefully shaped timbers selected from specific local trees. Women (and children) also played a role in the process. They sometimes harvested the wheat alongside the men, but more typically they followed the harvesters, weeding the land behind them and feeding the weeds to their livestock. In the final operations of all, turning the wheat into bread, the women ground the grain by hand-revolving one millstone mounted over another and feeding the grain between them. The women also most skilfully made the paper thin discs of local bread that were characteristic of the country.

Each man who contributed to the production and harvesting of the crop was paid according to a time-honoured formula. For example, the irrigators received one-fifth of the harvest, whilst the ploughman was given wheat equal to the amount sown and the harvesters shared one-tenth of the crop. So the income from the crop was very widely distributed.

Nothing of the crop was wasted. The grain, of course, was milled into flour for bread. The straw was fed to the animals or used for bedding so that when mixed with faeces and urine it formed manure which was fed back into the land. The chaff was mixed into clay to make either mud bricks or plaster. The stubble was grazed by the goats.

So, wheat exhibited all the virtues of our idealised rural community, but the system of production was clearly vulnerable as soon as more remunerative work became available. For example, the men who used the *graz* to make the irrigation basins earned, in 1974/5, RO1.33 per man day - equivalent to a mere £2.00. Naturally, almost any other job would have looked more attractive. But the action which effectively killed interest in wheat growing was the opening of the Oman Flour Mills in 1976 and the importation of bulk quantities of Australian wheat. In the mid-1970s the price of locally grown wheat in village markets in the Dhahira was RO17 per bag of 64kg (40 *mikyal*). If we assume a one hectare field, and a high average production of 2ton/ha, then the sale value was RO527. However, after payment had been made for seed and water, and everyone from the ploughman to the harvester had received his share, the net income to the farmer was only RO70. Meanwhile, as Australian wheat was becoming more readily available in the local market its price was falling, during the mid-1970s, from the equivalent of RO14 to RO10.6 per 64kg. This dragged down the price of local grain so there was no longer any profit for the farmer. Not surprisingly, therefore, he lost interest in growing wheat, and by the end of the 1970s almost none was grown (Dutton, 1982c).

In theory, and perhaps in practice if there had been sufficient time, the local system of control could have responded before the situation became critical. It could for example have: (a) found ways to reduce the labour requirement per hectare of wheat harvested, (b) rationalised the use of irrigation water and fertilisers, (c) as a result of reducing labour and water demands per hectare, increased the total wheat area, and (d) rationalised land tenure and field size in order to allow appropriate mechanisation and markedly reduce costs per unit of production. But the system of local control was neither knowledgeable enough nor responsive enough to begin to make the necessary changes listed above, and perhaps the attempt would have failed in any case given the fact that wheat lends itself so readily to mass, cheap production and bulk transport in and from countries such as Australia.

But the failure to respond has broken the communal bonds of interdependence that wheat cultivation fostered. Worse still, no alternative use has yet been found for the land on which wheat used to be grown. Higher value crops cannot tolerate being irrigated every eight days or less frequently as the *falaj* cycles through the village, so *falaj* land and water, both very precious resources in an arid country, are being used inefficiently or not at all (Dutton, 1986b).

Alfalfa was the other important minor crop grown in Oman, though unlike wheat it was grown both on the Batina and in the interior (Dutton, 1982c). In contrast with wheat, which did not survive the 1970s, as has been shown above, alfalfa not only survived the profound changes that were taking place in agriculture during the decade, but even increased in area. There are several reasons for this relative success of alfalfa, reasons which offset the fact that alfalfa, like wheat, was cultivated with the use of labour intensive techniques which, in fact, have remained essentially unchanged up to the present time.

In 1973/4 almost half the surveyed Batina gardens under cultivation were growing alfalfa whilst in the Ibri well gardens almost 90% had alfalfa and 100% of the relatively well-established gardens there were growing the crop. The continuing, and growing, interest in alfalfa on the Batina is reflected in the distribution of the crop between the old and the new gardens. Little or no alfalfa was found in the old coastal gardens but away from the coast the larger palm gardens typically had areas of alfalfa ranging from none up to 0.2ha squeezed between the palm trees. However, the largest areas of alfalfa (over half a hectare in four cases) were almost always found in the new gardens on the inland edge of the cultivated area. These gardens were being newly brought into cultivation at a time when there was a smaller demand for dates but a growing demand for fodder and vegetable crops. By 1974/5, only one year later, the area of alfalfa in the Batina gardens surveyed had already risen by 19.5%, and by 17% in the Ibri well gardens. Although the area under alfalfa has not been monitored since then there is no doubt that it has continued to grow.

Yet alfalfa, like wheat and date palms, was also very labour intensive, particularly in the creation of the basins used for flood irrigating the crop. Generally throughout northern Oman all the alfalfa was grown in *gelbas*, similar to those used for growing wheat. The length and breadth of the *gelbas* could in theory be adjusted according to the slope of the land, the rate of water flow, and the permeability of the soil, in order to obtain an even distribution of the water. But although there was a great range of *gelba* sizes they were not related to any practically useful criteria, and the long axis often lay down the slope so that when flooded the water was distinctly deeper at one end than the other.

The small size of the majority of the *gelbas* was almost certainly determined by the traditional method of irrigation whereby water was raised from the wells by the *zagira* trip bucket. This technique was very slow and the only way to ensure an even water distribution within the *gelbas* was to make them very small. The general effect of this was that most *gelbas* were smaller than they needed to be in the 1970s, by which time almost everyone had replaced animal power with diesel pump-sets. The small size of the *gelbas* had a number of disadvantages. First, it was a longer and more costly operation to make the *gelbas* and to provide them with an adequate network of irrigation channels. Second, more land was wasted in ridges and channels, and water was lost unproductively by evaporation from those ridges and channels. Third, there were more crop edges where weed infestation could start. Fourth, salts accumulated on the ridges as a result of water evaporation and then, after rain, were washed into the edge of the basin where they killed the plants and made weed infestation easier. Fifth, any possibility of mechanising the different phases of cultivation was eliminated, and finally, irrigation required the almost constant attention of one man to open and close the gates of successive *gelbas*. In summary, the small basins wasted land, water and time, inhibited the process of modernising the means of cultivation and significantly reduced production. Wastage of land is the key. Detailed measurements were made in contrasting plots of *gelbas* and revealed that in a plot with good, average layout and a simple network of channels as much as one-third of the land was occupied by channels and ridges, therefore one-third of the cultivated area was uncropped, with a proportional loss of yield potential, and one-third of the total evapotranspiration water losses were wasteful. It was calculated that by increasing *gelba* size from $15m^2$ to $50m^2$ the proportion of cropped land rose from 65% to 80%. Doubling the *gelba* size to $100m^2$ only increased the cropped area by an additional 5% but it had the important advantage of reducing the total ridge and channel length by 25%, with an equivalent saving of time in their construction and maintenance. Yet small basins (of about $20-2m^2$) continued to be used for growing alfalfa through the 1970s, and they remain the standard for alfalfa production today.

A brief description of *gelba* preparation will illustrate just how laborious the process was, and remains today. The land is first ploughed. Traditionally this work was undertaken with the use of a metal-tipped, single furrow, wooden plough pulled by either one or two bulls. The land, as for wheat, was ploughed twice, at right angles, the whole operation taking about 40 hours/ha. For some years during the 1970s bulls and extension centre tractors were alternative sources of ploughing power. The subsidised tractors were extremely cheap to the farmers, but the bulls remained in use because there was always a long waiting list for the extension centre tractor and because some of the areas to be ploughed were so small and constricted that the tractor ploughed very inefficiently, as mentioned above. After ploughing, the *gelbas* are made. The size and layout of the plot are initially decided upon, then the principal water channels aligned and shaped and their slope tested with water. The outer ridges of the plot are typically made first, perhaps aligned with the use of a string, then the cross-ridges are made. Gates to allow the water to flow into the *gelbas* are cut through the ridges, and the earth in the floor of the *gelbas* is smoothed. To prevent the soil being eroded from the narrow gates leading from the channels, the sides and bottoms of the gates are sometimes lined with palm fibre or other materials.

The commonest and most suitable tool used for making the ridges in alfalfa fields is the same *graz* used in wheat fields. But for alfalfa the ridges are made more robust becuase it is a perennial crop. The *graz* action is often done twice from each side of the ridge leaving marked depressions near its foot. In between successive scrapes the ridge is trodden by the men in order to make it firmer and resistant to water seepage. Finally, the depressions at the foot of each ridge are filled when the floors of the *gelbas* are smoothed. The only changes to this system being introduced during the 1970s were the gradual replacement of the locally made *graz* by imported pointed shovels for making the ridges and by rakes for smoothing the *gelba* floors and for making a fine seed bed.

It proved very difficult to obtain useful time-estimates for making *gelbas*; a day's work might start or end at quite different times, the number of men employed might vary from day to day or even hour to hour, and gaps of several days might be left between different parts of the operation. Therefore, only the briefest time-indications are available: two men took three hours (six man hours) to hand dig $70m^2$ and make five *gelbas*; two men took seven hours on $650m^2$ to make 30 *gelbas* excluding final earth smoothing; and two to three men took six days on $670m^2$ to prepare 40 *gelbas* including spreading dung and sowing the seed. This indicates something in the order of 300-600 man hours per hectare to make the basic *gelbas*, a huge time requirement, only feasible when labour is plentiful and cheap.

As far as time requirement is concerned alfalfa has the advantage over wheat because it is a perennial crop. However, this advantage is counterbalanced in three ways: because the *gelbas* have to last several years they must be made with more care; the crop has to be harvested 11-13 times each year instead of only once, when weeding has to be undertaken with rigour to prevent the alfalfa being infested with wild grasses during the summer months; and there was, in practice in the 1970s, a marked inertia to replant the alfalfa when it was old and weak. On the Batina it was not uncommon to see older *gelbas* of alfalfa which were thin and patchy and marked with white salt lines. Some cases were observed where *gelbas* were still being irrigated and cut although very few plants remained alive. Average yields could most easily and dramatically have been improved by ploughing in the older *gelbas* and resowing.

Irrigation was another time-consuming activity. Of ten timed measurements in Batina well gardens most ranged from 1.25-6.25 l/second and from 40-80 hours/hectare. These relatively slow pumpage rates demanded the presence of one man in the garden for the whole period, though he could often fit in some other work at the same time.

Harvesting, however, usually combined with weeding, was the slowest part of the operation. The alfalfa, harvested 11-13 times per year, was cut by hand. From one to three people - usually men, but boys, women and girls also did this work - might be found cutting alfalfa in a garden at any one time. Harvesting was done by hand, using a curved and serrated-edged knife, the *mingal*. A number of cuts were then bound together with a few twisted lengths of alfalfa to form either small or large bundles, the largest normally approximating to a *mann* weight or 4-5kg. Eighteen harvesting operations were observed and measured to reveal that on average it took 27 hours to cut one ton, and over a solid week (over seven days of 24 hours!) to cut one hectare. This is of course extremely slow but it must be remembered that under the system of cultivation, marketing and usage current up to the mid-1970s, whereby each farmer grew only a small area and sold only 1-2 *mann* to each buyer who made daily visits to the garden, the technique of harvesting was not inappropriate. Increasingly, however, during the late 1970s whole plots were being hand-cut in an afternoon for trucking to the market in Al-Ayn. The slow rate of harvesting was becoming a constraint on production but because no mechanical cutter was available, and because in any case it could not have worked within the small *gelbas*, the only way of speeding up the process was to employ 10 or more people (family members, friends and neighbours, and, on occasion, paid labour) working together.

As a result of the survey undertaken during 1973/4 and 1974/5 it was possible to give a fairly detailed estimate of gross income and profit from alfalfa. The estimate given here is for an established plot of alfalfa in the al-Khabura region; for a garden with both mean area and mean yield selling at the garden gate price current in 1974/5. It assumes 12 cuts per year, and also assumes that the farmer has the average number of livestock and feeds them the locally estimated mean of alfalfa per day. Therefore the farmer grew 0.22ha of alfalfa yielding 20.7 tons (at 94 ton/ha) of which 4.1 ton were fed to livestock leaving 16.6 ton for sale at a price of 82 baisa for 4kg, yielding a gross income of RO340. Set against this were irrigation costs (fuel and pump repairs) of about RO70 giving a net operating profit of about RO270 per year. But this example assumes that labour costs are nil (work being undertaken by the farmer and his family) and was, of course, only a small income for a very considerable amount of work. And in the year in which the *gelbas* were prepared and the alfalfa sown, normally involving labour costs of RO1-2 per man/day, the net income will have been much reduced which helps explain the reluctance of farmers to plough in old alfalfa and resow.

But given the fact that alfalfa was almost as labour intensive as wheat when grown in Oman why did it continue to increase in area during the 1970s, and 1980s, instead of declining to nothing? Perhaps the principal reason is that its price was not undercut by a cheap, mass-produced import equivalent to Australian wheat. In fact the alfalfa price to the Omani farmer rose quickly in the late 1970s as a result of a rapid growth in demand for the crop in the market in Al-Ayn. Per *mann* of freshly cut green alfalfa the price rose to 400-500 baisa. Although the costs of pump-sets, fuel, seed and transport had also risen, the high selling price made alfalfa a very attractive crop. Selling at 400 baisa/*mann*, a man with half a hectare producing 50 ton/year would gross RO5,000. No other crop whose husbandry was currently understood by a large number of small farmers compared with alfalfa in productivity, sale price and relative freedom from pests and diseases. However, the new market broke the traditional interdependence between forage producer and local livestock owner. Less forage was available on the local market and at a much higher price. Exports also meant that Oman was exporting an increasing amount of its scarce water resources, using neither the water nor the forage to the benefit of the local livestock industry. But it is interesting to note that alfalfa production increased substantially without the help of any major saving in labour requirement. Even today almost all alfalfa is grown in hand-made irrigation basins and is cut with hand-held serrated knives.

2.2.5 *Irrigation and water supply*
In rural communities in Oman by far the dominant usage of water was the irrigation of agricultural crops. From the beginning of our field research programme, this posed questions about declining water yields, particularly as the 1960s had seen the substitution of many *zagiras* by diesel pumps with a much higher potential extractive power (Letts, 1978b).

In well gardens attempts were made to gather data on the history of fluctuations in static water levels to give some idea of medium- and longer-term trends. Some information was found which indicated fluctuations in rainfall over the previous century. The evidence suggested an eleven to twelve year periodicity of droughts.

Clues about declining well water levels in the Dhahira during the decade or so up to the mid-1970s were obtained by asking farmers about the history of their wells since pump installation; especially the need to lower the pump platform, declining yields of water, and recent well deepening. Additional information was obtained by measuring the height above water of rings of calcium carbonate deposited on pump pipes, and from the number of additional lengths of rope added to scoop-hole buckets. In the villages of Tan'am and Dariz, in particular, evidence of falling water levels was common, partly because these two villages had the oldest pump wells and therefore had seen the greatest range in water levels. The sequence suggested a regular and progressive decline in water level of some 33cm/year over the previous decade, and most farmers reported a marked decrease in water supply. Although changes in other villages were less marked, of the 75% of wells older than two years for which data were obtained, 70% had been noticeably affected by increasing water shortage. In the decade up to 1973, and continuing on to 1975, there had been a general trend towards falling water tables and declining well yields. During the decade up to 1975, and at an increasing rate within that decade, new wells had been equipped with pumps and old *zagira* wells had been converted to pump wells. By the mid-1970s, pumps (medium lift and powered by diesel engines) were found throughout the whole of settled interior Oman, including the Dhahira. As has been stated: 'The causes and consequences of this increase were of fundamental importance to the water-resources, economy and society of a large region along the western side of the mountains' (Letts, 1978b, p. 66). On the Batina in the two decades or so prior to 1975 the *zagiras* had also almost all been replaced by pump wells but on the Batina no decline in static water level or in water yield was reported.

It is instructive to compare the output of pumps with that of *falajes* and *zagiras* (Letts, 1978b). First the comparison with *falajes*. It was noted that:

'During the winter of 1973/4, the 128 pump wells in the catchments of the wadis Al-Aridh and Al-Kabir extracted ... approximately 2.5mn gallons a day. The is almost exactly the same amount as the flow of ... the larger of the Ibri *aflaj*. [*falajes*]' (p. 80)

In 1975 the average cost to create and equip a new well in the Dhahira was RO1,500, so 128 new wells would have cost almost RO200,000. This was at a time when a well-digger earned only about RO2 per day so the total time requirement to create 128 wells would have been very high. In fact, the time and capital cost would have been of the same order of magnitude to construct a *falaj* of equivalent output and which would have had lower running costs and been less threatening to the water table. Comparison with the *zagiras* shows that:

'...the 128 pump wells in the sample produced 2.5 times the volume of water hypothetically extractable by 163 *zagira* wells [1.1mn gallons a day], and 15 to 20 times the probable actual *zagira* production [during the winter months]'(p. 81),

and they produced 2mn gallons a day in the summer months when, previously, the *zagiras* had not been used. The diesel pumps thus posed an important threat to the sustainable use of an essential and scarce resource, water.

In the study area around al-Khabura, it was calculated that in comparison with the previous *zagiras*:

'...the present extraction rate stands at just under 200% of the original winter rate and 180% of the summer rate. Thus the introduction of pumps has led to an approximate doubling of water extraction in this area of the Batina.' (p. 84)

Small farmers spent about RO250,000 to establish pump farming within the field area excluding the Batina coastal strip, and perhaps some RO400,000 in total. This growth of pump-wells in the field area occurred without any substantial encouragement by development companies or the government. As an expression of the willingness, even enthusiasm, of farmers to invest in agriculture the investment during the decade to 1975 was a very strong indicator of a long-term commitment to farming.

In some instances the change from *zagira* irrigation, which was extremely labour intensive for man and animal, to diesel pump irrigation, which required almost no human effort, led to gross inefficiencies in water use. For example, in two gardens at al-Khabura where the owners were both traders in the *suq*, they switched on their pumps, went to the *suq* and left the pumps running unattended for hours. In one case each row of 14 date palms was watered through one gate and it took 3.5 hours, with a flow of 6.5 l/s, for water to reach the last tree in the line. In the other case, the row was 35 palms long, and the final tree was only wetted after 5.5 hours with a flow rate of 10.7 l/s. In these cases the first palm received up to ten times as much water as the last.

In *falaj* settlements, however, it appeared that despite many vague claims of declining water yields, no evidence could be produced of such declines in the decade up to about 1973 (Letts, 1978a).

Tests on quality showed, in general, that the quality of *falaj* water was considerably better than well water. In consequence, and perhaps aided by the fact that *falaj* irrigation rates are higher (which automatically provides a surplus of water for leaching unwanted salts), *falaj* cultivation has continued with a minimum of crop rotation for hundreds of years without serious detrimental effect. In contrast, crops dependent on well water sometimes suffered salinity problems after only ten or twenty years.

Inspection of the *falajes* in 1974/5 showed that in many cases in the collecting area of the *falaj* upstream of the mother well, in the vicinity of the mother-spring, roof collapse was common, and there was need for maintenance. Regular maintenance of the whole *falaj* system is an on-going requirement, yet field evidence suggested that lack of maintenance in the previous decades had left many *falajes* in poor condition with much maintenance work needing to be done. Work was being undertaken, however, on the network of distribution channels. Some of them were being lined with cement whilst others were provided with diversion gates more watertight than before. Yet water losses both above the settlement and within the settlement remained high, ranging from 18-20% to 73-6% of the water as measured at source. Within the settlement most water loss was due to poor damming of the alternative routes, but there were also losses due to leakage and seepage. The losses were made worse by irrational patterns of water cycling. For example, a group of gardens along a channel might not be irrigated in succession but on different days within the village cycle.

Some *falaj* water, typically, is owned by the *falaj* itself, and proceeds from sales of the water to the farmers have historically been used for *falaj* maintenance and improvement. However, a typical *falaj* owns only about 10% of the water, and its income from this is small. Water may also be owned by members of shaikhly families. Where most of the water is owned in this way, the shaykhs rent out surplus water to people who own land but have no water rights. This benefits the shaykh but all too often, over time, he has lost any sense of obligation to use this money for the general upkeep of the *falaj*. Where shaikhly water ownership is dominant it was found that running repairs tended to be few and many palm owners had to mortgage their palms in order to rent water with which to irrigate them. At this time, as has already been shown, the relative value of the date harvest was falling as people were gaining access to a much wider variety of imported foods. Therefore the financial returns from dates to the *falaj* system failed to keep up with the growing costs of maintenance. In fact, during the 1970s MAF took financial and technical responsibility for maintaining the system. This was not sustainable long-term and, worse still, diminished the community sense of responsibility for repairs and extensions to the system. Little or no attempt was made to incorporate other, higher value crops into the production cycle, crops which unlike dates required the high quality water typical of the *falaj* and which would have provided a higher financial return to give the system a secure future. Thus the *falaj* system, complete with its inherent conservationist strengths, was being undermined by sinking an ever-growing number of wells equipped with diesel pumps. Perhaps the most unfortunate consequence of this deterioration of the system was the progressive weakening of the *falaj* committees which traditionally played an important role in water management. These committees needed to be strengthened and integrated into a national water management structure. The self-reliance and inter-dependence which they conferred onto the *falaj* communities was in decay.

The Batina gardens were a special problem area because, in the short to medium term, water, apparently, was not a limiting factor in crop expansion. A comparison of all crops showed that date palms, per unit area, were, in practice, the least demanding of water. The significance of this is that with the rapid decline in the value of the Batina date palms and an increase in interest in growing alfalfa (later Rhodes grass) and a wider range of vegetables, a switch away from date palms, as in the new gardens at al-Khabura, was leading to a rapid growth in overall water consumption. In the mid-1970s at al-Khabura the palms, which occupied over 80% of the cropped area, used only 47% of the water, and although it is clear that this represented under-watering it is equally clear that a significant move from date palms to vegetables and forage crops would greatly increase the water demand, a demand not checked by any awareness by the farmers that high levels of consumption would lead longer-term to falling water tables and poorer water quality.

2.2.6 *Handicrafts*
The wide range of handicrafts traditionally practised in rural Oman employed the manual skills of an important number of men and women to make goods which played various roles central to the economy and society of village and pastoral life. Handicraft production was normally small-scale, often single people working by themselves or with members of their own families. However, in Oman as a whole total production of some particular handicrafts played a significant role within the overall rural economy because practitioners were to be found in each village or other type of rural community.

Handicrafts utilised natural physical resources together with the produce of agriculture and fisheries, and sometimes imported manufactured goods (such as old car springs) which were reworked to produce new manufactured items. Handicraft skills were gradually evolving so that side by side with inherited skills were new skills acquired to process newly available raw materials or to work with newly available tools.

Before the oil era a very wide range of handicrafts was practised, and an indication of their main categories was listed by Dutton (1983a), classified by material used. The list illustrates the everyday role that most of the products played in the rural economy and domestic life and it also shows the range of raw materials, both local and imported, that were employed. As primary sources of craft materials the local date palms, livestock and clay were of particular importance, though imported metals and fabrics had long been essential as well.

Some of the work was undertaken by full-time specialists, including weavers, silversmiths and potters, but other craft activities were practised on a part-time or domestic basis by a large number of both men and women. The latter skills included spinning, the manufacture of palm-frond products and the processing of milk. Some skills were practised by specialists and non-specialists alike:

weaving and the tanning and sewing of leather are good examples - the larger villages had weavers and a tanner (*shammar*), but members of most pastoralist households knew how to tan leather and sew milk churns and many women in the households knew how to weave.

As a source of craft materials the date palm is of particular interest in that most parts of the palm can be utilised for craft work, to make a wide range of important goods. As palms occupied 70-80% of the cropped land, many rural families were skilled in making one or other of the palm products.

Apart from dates, the palm yields fronds with a long, straight midrib, a butt, long, tough leaflets and sheets of fibre. The fronds were used in a variety of ways (as described in more detail above): primarily they were joined side by side to make *barusti* sheets used for constructing the walls and roofs of houses; and secondly, they were tied together to form the hull and deck of the *shasha* float-boat.

Rope made from palm fibre was used to bind together the fronds which formed the hull of the float-boat. The same rope was also used in the manufacture of *barusti* sheets and in a hundred other ways. From the palm leaflets were plaited endless braids which were subsequently sewn edge to edge to make a very wide range of mats, baskets and bags. The palm trunks themselves were skilfully split to provide timbers for house construction and, in the Rustaq region, lengths of trunk were hollowed out to form beehives.

Omani livestock produced (in addition to meat) skins, wool, hair and milk, all of which have formed the raw material of other craft skills. Skins have been tanned according to complex and locally evolved techniques using the pods of the tree, *Acacia nilotica* (*qarata*), as a tannin source. For the farmers the tanned skins were fashioned into the huge leather buckets (*dellu*) used for raising irrigation water. For the owners of cows, other skins were sewn into milk churns and for the pastoralists and Bedouin, leather water containers were made. For more general usage, sheep skin rugs, sandals and belts, etc. were also manufactured. As for the wool and hair, this had long been spun and woven by both men and woven into a wide range of products including tents, bags and donkey and camel trappings and harness. Milk had been treated and churned to produce both *leben* and butter.

Other Omani crafts were related to Oman's use of the sea. First amongst these is boat building. All types of craft from large ocean-going trading ships to the float-boats mentioned above had been made in various coastal villages. Additionally, traditional and modern forms of fishing traps were made, and fish had long been preserved by skilled drying and salting.

Other crafts were dependent on local natural resources. First, a number of regional potteries made a very wide variety of vessels, including beakers, water-cooling jars, large water-storage jars up to a metre high and pots for storing dates. Second, carpenters in most villages were skilled at using the local wild trees in the manufacture of doors, window frames, ox ploughs, well-head gear and the framework of the plank boats. Also, a number of people were skilled in such crafts as making gypsum cement and in making and using paints from local ores. Dyeing yarn was also practised, using locally grown indigo or other plants.

In addition to the above, other craftsmen worked with imported raw materials, notably the weavers of cotton cloth, who used imported Indian yarn, and the metal-smiths, who made copper vessels and silver and gold jewellery. Village blacksmiths working with old car springs and other scrap metal were responsible for the manufacture of tools used by the farmers, including plough shares, sickles, hoes, diggers and palm trimming knives.

The crafts enumerated above were all essentially functional in character; they were not for tourists, nor did any of them evolve into quality work of high aesthetic value, such as the knotted carpets of Iran. However, a few of the more intricate camel trappings and one or two of the products of the silversmiths might be considered exceptions to this rule, notably the decoration of the *khanjars* (curved daggers) worn by the wealthier men on ceremonial occasions. Normally, however, the craftsmen provided the range of basic goods which the local communities aspired to and which were of key importance for the primary producers and for domestic life.

The Bedouin, with their mobile life style, had relatively little need or use for material goods. However, most rural communities in Oman were settled in villages and always had much higher material aspirations than the Bedouin. The village people also invested much more time and effort to ensure the continuing realisation of their individual and communal material aspirations. And the fact that the village communities were settled in one place not only permitted the acquisition of a far greater quantity and range of material goods but allowed the time and opportunity to develop the necessary skills to acquire them, so such goods were mainly provided for the community by those of its members with handicraft skills. Craftsmen played a very important role in making possible the characteristics of a rural community, stressed in this volume, of mutual dependence and collective self-reliance, by creating the multiple linkages in the social networks which are one of the basic elements of community structure. For example, the craftsmen linked the farmers with the fishermen in the making of the *shasha* float-boats from palm fronds; they linked the producer to the consumer in, for example, the supply of dried fish; they linked the farmer with the available raw materials of his environment through the manufacture of hand tools, ploughs and well-head structures required for the traditional bull-operated *zagiras*; they linked the farmer to other farmers and to householders (and even to beekeepers) through the multiple craft use of the date palm and the by-products of sheep and goats; and they linked the pastoralist to the villager through the same range of livestock by-products. They also made the householder intimately dependent on his physical environment by providing from it the bricks, plaster, gypsum cement and paint used in house construction, as well as the timbers, mats and baskets etc. manufactured from the date palm. Additionally, the craftsmen provided some goods for sale outside the village for the purchase of items which could not be produced locally. They also manufactured some goods from imported materials including the *dishdasha* sewn from imported cloth, and the *wazar* (worn by all men under the *dishdasha*) woven from imported and local yarn.

Therefore the practice of craft skills created a range of mutually respected and skilled people who were essential to the survival of the community. The employment and handing on of these skills also created the sense of interdependence, self reliance and local control which, it has been argued above, played an important role in creating social cohesion, and they preserved a valuable amount of independence from the outside world and undesirable changes in it. Additionally they minimised the need for imports by fully utilising local materials while at the same time creating some exports. The crafts also provided skilled work for the old and infirm thereby giving them an active role within the community.

But the oil industry in the Gulf brought big changes to Oman's craft industries even before oil was discovered in Oman itself. Men were attracted from Oman to the Gulf states, lured by the prospect of regular work for a relatively high and dependable salary. Once in the Gulf the migrant labourers with their new money had access to a wide range of imported goods that they had never seen before. Their material aspirations rose and they began purchasing imported goods for use at home including some whose equivalent had previously been craft-manufactured in Oman. Thus the craft industries suffered in two ways. First, as craft work in Oman was very labour-intensive but yielded only a small net income, jobs in the Gulf seemed particularly attractive to the craftsmen. In many cases the sons of craftsmen took the new jobs and so either never learned or never practised their inherited skills. Second, cheap imported rugs, mats, plastic cups and containers and many other products all prevented the price of their locally manufactured equivalent from rising. Thus the craftsmen could not increase their income even to counter the effects of inflation, much less to enable them to meet their own enhanced material aspirations. The oil industries in the Gulf and finally in Oman itself engendered much rural decline but within the spectrum of rural decline the craft industries suffered most. Both agriculture and fisheries were larger industries with greater resilience and given more support from government authorities; Omani rural crafts were not considered worthy of much attention by Oman's urban population. Crafts also suffered heavily because they were particularly labour-intensive; a weaver, for example, used to work a very long day for only a small reward. Although other rural products such as fish, alfalfa and goats increased in price to reflect the new regional wealth, this was not possible for the craft products because, as mentioned above, they were heavily undercut by cheap imports. Crafts also suffered as an indirect result of the speed of change from a pre-oil to an oil economy. This gave no opportunity to the craftsmen to modify either their technical processes or their final products to fit new market requirements and opportunities. Skilled craftsmen were being reduced in status to unskilled labourers and the community role that they had performed in building the self-reliance and inter-dependence of the villagers was almost lost. Craftswomen also were similarly being disinherited, particularly those women *shawawi* and Bedouin who span and wove. Additionally, the role in craft work for the old and infirm was being destroyed. The decline of the crafts provides a poignant example of the way that under the indirect influence of oil wealth the Omani villages were moving away from a state of being developed, as defined at the outset of this chapter, to one of being undeveloped.

With the decline in crafts there was also less use of the raw materials provided from the local environment and therefore less understanding of them and less awareness of the need to preserve them. In addition, the main agricultural products, date palms and livestock, had less overall value because their by-products were no longer utilised.

2.2.7 Fisheries

Fishermen operated from villages throughout the length of the Batina coast, including those villages around al-Khabura. In 1948 Bertram estimated the number of adult fishermen on the Batina at 'about 25,000 or less'. Even if the figure was indeed less than 25,000 there would appear to have been a significant drop in numbers by 1971 when Whitehead estimated a total of 15,000, and by 1972-3 when Mardela's survey points to an even lower total (Donaldson, 1978a). In 1978, Bertram (personal communication), on a visit to al-Khabura, contrasted the relatively small number of *shashas* to be seen on the beaches then with the same beaches that had been 'covered in *shashas*' in 1948 (Figure 3).

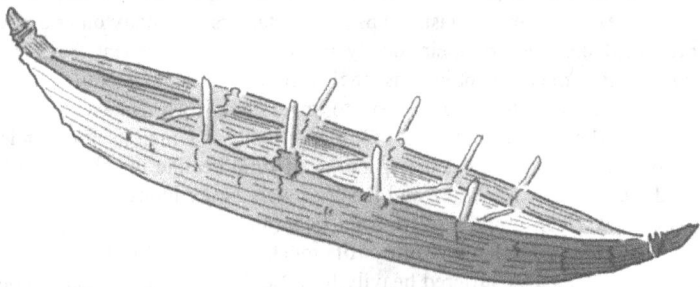

Source: Donaldson, 1978

Figure 3. Shasha or 'float-boat'

The very careful survey undertaken by Donaldson (1978a) in the mid-1970s suggested, in fact, that the total number of fishermen on the Batina stood between 3,500 and 3,600 only. What had happened to the others? In the 1970s surveys it was found that half the fishermen respondents also owned gardens, and even though few of the remaining fishermen regarded the gardens as economically significant, some ex-fishermen may have become farmers. However, fishermen who supplemented their traditional fishing incomes with extra work, either locally, elsewhere in Oman or in the Gulf states, almost all undertook the extra work within the fisheries sector, mostly fishing but some fish trading. About 40% worked periodically in the Gulf, usually Bahrain, as captains or crew of boats owned by fishing companies. Thereby, the fishermen enhanced their fishing skills, became familiar with a wide range of new boats, techniques and equipment that the companies were introducing in the Gulf, and earned sufficient money to buy similar equipment which they took back to their home villages on the Batina. The new equipment reduced the labour input and will have increased productivity, and the fishermen were able to charge higher prices for their fish because, in contrast with the craftsmen, the cost of their produce was not being undercut by cheap imported equivalents. Indeed, one reason for high prices was that some fish were being exported (mixed with ice in tanks on pick-up trucks) to the Emirates and even Saudi Arabia where, because of fish shortage and oil wealth, the price of fish was very high.

The introduction of petrol and diesel engines to the fishing industry began in the early 1960s, and became generalised during the 1970s, roughly contemporaneously with the introduction of diesel pump-sets in farming. But whereas pumps undermined the role of traditional artisans and created a skill-gap in engine repair and maintenance which, all too often, was filled by immigrant workers, and eventually became a threat to the quantity and quality of water upon which agriculture depended, the introduction of boat engines neither undermined traditional skills nor threatened the long-term future of the industry, though it did create a skill-gap which caused long delays in engine repairs in the 1970s.

In fact, new fishing equipment was introduced in a manner and at a rate which allowed it to be integrated smoothly into the indigenous system. Fish traps are a case in point. The original traps, made of date fronds, were heavy and cumbersome to handle and their size was limited by weight (they absorbed water) and by frond length. They were replaced during the decade up to the mid-1970s by wire traps. Although the wire was imported, the traps were still locally made and repaired, and they had the advantage of being lighter to use even when larger, and if the wire had been coated they lasted longer than the date frond traps. Fish nets afford another example. During the late 1960s cotton nets were replaced by nylon. Pieces of netting were bought in Dubai or Mutrah, but the fishermen made up the pieces into the patterns required for different nets, and added ropes, floats, weights and anchors. The fishermen also repaired the nets. Finally, a most interesting case is the manner in which the outboard engine was adapted to the *shasha*. The *shasha*, described above, might be regarded as the most characteristically 'primitive' of the local fishing vessels and at first sight it would seem impossible to motorise. But in January 1975 the first *shasha* was seen near al-Khabura with an outboard engine, as described above. A rectangular hole had been cut through the body of the *shasha* towards the stern and a board secured to the forward edge of the hole on which the engine had been mounted. By August 1976 on the same stretch of coast 12 motorised *shashas* were known, and the number and proportion of them increased in subsequent years. The boat builders gradually and very skilfully adapted the design of the *shasha* to match the power of the engine by making it both longer and broader, so the engines gave the *shashas* a new lease of life (Dutton, 1987d).

So, the fishing community though reduced in size in the years up to the mid-1970s had itself integrated new engines and nets and other fishing equipment and techniques into the traditional fishing system and thereby made the industry less labour demanding and more productive. The fishermen had also developed, for themselves, new market opportunities in Oman and in other Gulf countries. In consequence, although the fishermen were fewer in number the industry remained viable, and the fishing community was fairly successfully shifting the dynamic balance whereby they were able to take more from the system as a result of higher investment to cater for present needs and enhanced aspirations without threatening the industry's longer-term future.

Nevertheless, some interesting specialist skills were being swept away, mostly as the result of the introduction (with government aid) of refrigeration units and ice plants. Preserving fish always was, of course, a difficult problem in a country with a climate as harsh as Oman's. Yet without the ability to preserve the catch it would have been impossible to exploit the seasonal occurrence of big shoals of either the small fish (sardines and anchovies) or the large *scombridae* such as the kingfish. So, local techniques of preservation had been developed to meet the need, and chief amongst the techniques were wet salting, dry salting and drying (Donaldson, 1978a). Wet salting was practised mainly with the *scombridae*, though other medium to large fish with the exception of sharks were also treated in this way. The fish were split along their underside and the viscera and gills removed. Whole fish were then opened out flat (very large fish were cut into pieces), the flesh scored deeply down to the skin, and powdered rock salt was then rubbed generously into the flesh. Fish were thereby preserved for 2-3 months or more. Dry salting was reserved for the treatment of sharks and rays only. All the stages of the wet salting technique were followed, and the salted flesh was then left to dry in the sun. The pieces of dried shark (known as *'uwal*), hard but slightly pliable, were both light weight and very durable, characteristics which made them particularly portable and so suitable for transporting into and within interior Oman where the population did not share the coastal dwellers repugnance for shark, which is thought of as a scavenging animal. Drying without salting was (and remains) practised with sardines and anchovies and, unlike the previous two methods, was carried out on the Batina by the fishermen themselves rather than by the fish traders. The technique was simple; the fish were spread thinly over a prepared drying floor on the upper beach, sometimes protected with crude fencing to keep out animals, and raked periodically for 2-4 days while drying in the sun (Donaldson, 1978a).

Up until 1970 the lack of motor transport was the major constraint on the sale of fresh fish in any part of Oman other than the coastal fringe of a few kilometres width. Fresh fish was unknown in the interior. But with the change of regime in 1970, and the relaxation of prohibitions on transport, it was found that the people of the interior liked fresh fish quite as much as people on the coast, and were prepared to pay almost twice as much for it. As is often the case, the new and strong demand produced the entrepreneurs capable of exploiting the new opportunity for sales. Many of the new fresh fish traders-transporters had previously traded in cured fish. To carry the fresh fish they invested in pick-up trucks on which were mounted sheet aluminium and plywood boxes filled with a mixture of fish and ice, and insulated with polystyrene. But although one might regret the loss of techniques of fish preservation dependent on renewable solar energy and their replacement by techniques based upon non-renewable fossil fuels, at least the new methods had been incorporated into the fishing community, a fact ultimately made possible because the restructured fish catching and marketing system remained profitable.

2.2.8 Marketing

A general but important conclusion from the Durham University marketing study was that merchant and entrepreneurial activity was extraordinarily strong almost everywhere in Oman (Donaldson, 1978b). The vitality and flexibility of market networks and their widespread ramifications coupled with the long-maintained attractiveness of trading as an occupation have important implications for development. For most of Oman's history, long-range mobility has been a normal feature of life for the people of northern Oman, including both maritime and overland trade links. This mobility and these links had been greatly strengthened by economic and technical changes within Oman and in its neighbouring states. In consequence, the consumer market in Oman was, in attitude, remarkably open to import penetration. With a few exceptions (such as a preference for local honey and for the local goat) it was considered normal for commodities to be available from the whole world rather than merely from the local region, a trend greatly reinforced by the rapidly growing cash flow from remittance money earned in neighbouring countries where imported goods were commonplace. Within Oman the pastoral Bedouin, the *shawawi* of the mountains and plains, the *falaj* oasis communities of the interior, the Batina coastal settlements and the regional and national urban centres were, and for long had been, interdependent through trade. To these networks had been added a growing strength and diversity of international trading links, particularly since the mid-1960s, helped by new roads, more vehicles, freer international travel and a vast increase in disposable wealth. By 1978 the value of goods entering Ibri *suq* was estimated at RO2mn annually, and RO1.2mn in al-Khabura *suq*. Of these supplies 90% were imported in Ibri and 80% in al-Khabura (the major local product being fish).

During the early 1970s al-Khabura *suq* experienced a period of great and rapid change, expressed most obviously in the building of new shops and the renovation and rebuilding of old ones, together with the ever-widening range of goods sold in those shops. Rents and prices were also rising rapidly. The period of most rapid expansion in Ibri *suq* occurred rather earlier than at al-Khabura.

In Ibri *suq* the main non-imported goods were dried dates and fresh dates in season, live animals, fresh, local fruit and vegetables, and, from rather further afield, fresh, dried and salted fish. Locally gathered firewood and locally produced wood, charcoal and pottery were also sold by auction in the *suq* in varying amounts, but neither the quantities nor the values involved were large. Locally produced alfalfa was also on sale.

Like fishing, rural trading was also adapting very effectively to new opportunities that were being explored (sometimes unsuccessfully) by large numbers of would-be entrepreneurs without the assistance of either planning or feasibility studies. The dynamic balance shifted rapidly and radically to attain new equilibria with greatly increased throughput and profits matched by ever-growing investment.

2.2.9 *The disintegrating community*
The Durham surveys revealed, amongst other things, the extent to which the traditional rural communities were being affected by the impact of oil. In particular, a high proportion of the most active age group of men was employed outside their villages, often outside Oman in the Gulf States. Only rarely did they use their inherited artisanal skills and in consequence skilled and experienced rural producers were being turned into a raw and largely unskilled labour force. Any new skills they learned, in the army or police force or the oil industry, were mostly irrelevant to increasing the productive capacity of their villages. The absence of the men from home effectively limited the villages' productive potential. An increasingly high percentage of village wealth was being created in the form of salaries earned elsewhere. The men who earned these salaries spent an increasing proportion of them on goods manufactured overseas. On the one hand these goods included items like pump-sets, which showed a willingness to invest in the future of agriculture, but on the other hand the list of new purchases included goods such as plastic sandals, woollen rugs, cement for concrete blocks, timber, plywood, brushes, baskets, rope, beakers, water pots, storage jars and small agricultural implements which all replaced goods that had been manufactured at home. This kind of action wasted local raw materials, undermined the position of the skilled craftsmen, ended the interdependence of the members of the community and made the community as

a whole unnecessarily dependent on the outside world. It had, by the mid-1970s, almost crushed Oman's small but previously well-established craft manufacturing industries and also, in the case of things like rope and mat making, reduced the opportunities for the old and disabled to be active and self-sufficient members of the village community.

Major sectors of the agricultural industry were also in decline. The date palm, easily the dominant crop, lost value as a foodstuff and as a source of craft materials. Wheat, an important minor crop in the interior villages, disappeared almost completely from the *falaj* gardens and could not be replaced because no one seemed to find the time or incentive to adapt the eight-day *falaj* water delivery cycle to the on-demand needs of high value crops. The dried limes produced on the Batina coast lost their export markets in the Gulf with no effective attempt being made to replace them. The various branches of livestock production (Bedouin, *shawawi* and village) were in decline, and poor productivity nevertheless entailed over-usage of range resources. Government support, in the form for example of highly subsidised extension services and free repair and extension of *falaj* galleries, served mainly to undermine traditional self-reliance whilst inhibiting entrepreneurial advances. Also, the rapid switch from animal to diesel power to lift water from the wells for irrigation led to water being used with ever greater inefficiency. Finally, in agriculture, much of the indigenous labour force was replaced (during the late 1970s, and 1980s) by cheap and largely unskilled labour from the Indian sub-continent.

In brief, the surveys revealed that the indigenous craft and agricultural communities were in disarray. Gone was the feeling of mutual dependence on each others complementary skills. Gone also, in an age when everything could apparently be provided by oil wealth, was the sense of obligation to the future. Gone, therefore, was the concept of living in harmony with the environment and going was the independence and the skilled use of local resources that was associated with it.

But contrast the above with changes taking place in the fishing industry, in marketing and even in one branch of agriculture - the production of alfalfa. These traditional economic activities had shown themselves to be much more flexible in adapting to changing circumstances. One factor (in fishing and alfalfa growing) was the absence of competition from cheap imports. Another (fishing and to some extent marketing and alfalfa production) was protection from cheap, foreign labour. Another was the relative lack of government interference. For a combination of these reasons, the small-scale private sector in the fishing industry had shown itself fully able to change almost every aspect of its system, and raise levels of production and income for those who stayed in the industry.

Meanwhile the government was increasingly taking responsibility for a wide range of village activities. In many ways this was beneficial and necessary - e.g. the establishment of schools, extension centres, hospitals and clinics - but it had three weaknesses from the point of view of development as we have defined it. First, national organisations were sometimes too complex and remote for them to be fully aware of or responsive to local requirements. Second, it was reducing people's local sense of responsibility for their own future. Third, it diminished the self-reliance of the rural community by weakening that measure of responsive local control over the development equilibrium which had depended on fear of failure. Furthermore, not only was the concept of working in harmony with the environment weakened but the idea that responsive control over resource management was necessary was tacitly challenged in a period when everything that anybody wanted arrived on an apparently endless stream of government-provided oil wealth.

Did the disarray in the craft and agricultural communities matter? The answer to this question is yes, for two reasons. First, the stream of oil wealth is not endless. Oman itself is not a particularly oil-rich country and all but the wealthiest of its immediate neighbours will only be rich in oil for a few more decades. Therefore Omanis must look to a future when they are again economically dependent on the produce of the land and the sea and on their own skills. Thus, the skills needed for work in their homeland must be developed, not allowed to disappear. Second, as rural Oman appeared ever more irrelevant to the national economy, the social and spiritual life inextricably associated with rural communities had begun to break down. The consequences of this were multiple, and one of the worst, rural-urban drift, was predictable because it had already characterised most Middle Eastern countries undergoing rapid change during this century (Clarke and Fisher, 1972). In most cases this population movement is associated with a wide range of undesirable consequences, though also with benefits.

In consequence there was an undoubted need for rural community development in Oman, a need which the government of Oman was resolved to tackle. Based on the experience gained during the survey programme we were privileged to be asked, through the implementation of the Khabura Project, to contribute to this programme in the central part of the Batina coast, Oman's major agricultural region.

It was the awareness that the rural communities were being dislocated, coupled with an equal conviction of the need to rebuild them that gave shape to the conceptual framework upon which the aims and activities of the Khabura Project were based. This did not imply any desire to reconstruct the villages as they had been previously. A new and more productive dynamic balance had to be established by making appropriate use of new technologies and methodologies. The Project's activities also had to make every effort to keep in step with the changing context of government policy and to adjust to the improving range and quality of services which the government was providing for rural areas. At the same time we had to create opportunities to inform policy and planning processes whenever our first-hand field experience gave us particular insights into actions or approaches that were more likely to be beneficial than others.

3. The Khabura Project: Design

3.1 Guiding concepts

At the outset it was stated that the general objective of the proposed Project 'is to achieve a self-energising movement towards higher living standards' (Bowen-Jones, 1974a). The higher living standards were to include both economic and social desiderata, and the project was to be based within a given Omani village community. Other objectives included training in aspects of rural development and 'field-testing' technical and other innovations (Bowen-Jones, 1974b).

The conceptual objectives were that the project should assist people to become: less dependent on the outside world (with an increasing use of and reliance upon local resources); more mutually self-reliant (with a dependence on the complementary activities of the members of the community); and more aware of future needs and the obligation, therefore, to live in harmony with their environment.

At the same time there had to be a realistic acceptance of the new, oil-based economy which had given people higher aspirations and made them potentially part of a wider national community. Oil had helped create the problems defined in Chapter 2, but it also created an opportunity to set the equilibrium of economic activity at a higher level so that some of the new personal and community aspirations could be satisfied. But if, thereby, more was to be taken out of the system in order to satisfy these aspirations, more would have to be put back into it to achieve effective and stable growth. In a region with virtually no virgin resources this could only imply the application of longer hours, harder work, a wider range of remunerative skills, or more productive labour.

But were these aims realistic in Oman in the mid-1970s? Certainly, objectives that simply required longer hours and harder work would have been unrealistic. The traditional organisation of rural activities in Oman was already dependent on long hours of hard manual work in a harsh and enervating climate, an unattractive prospect given other opportunities. But just because production had been so dependent on manual labour it was regarded as possible to replace some labour inputs with improved techniques, without becoming over-reliant on high-cost alien technology. Therefore movement towards more productive labour was of prime importance to project strategy.

Because of emigration to work, the number of active local producers had diminished, but, if those that remained showed that they were responsive to demonstrable opportunities for change, there was a point of departure. Such 'enterprising peasants' had been found in Africa (Hill, 1970; Tiffen, 1976) and in other parts of the world (Schultz, 1964). They could also be found in Oman, as the research surveys had already demonstrated. For example, by the mid-1970s nearly all the farmers who had previously lifted water using a *zagira* had replaced this with a diesel pump-set. During the mid-1970s, some of the farmers also responded to an expanding market for alfalfa in Al Ayn by increasing their alfalfa acreage. Meanwhile the fishermen had replaced their cotton nets with nylon nets and they were using new types of fish traps and fitting their boats with inboard and outboard engines (Donaldson, 1978a, 1981). The fishermen, as has been shown, had adjusted to the new economic situation better than other groups, but because their numbers had fallen they were catching fewer fish than before. Craftsmen, meanwhile, had made no changes at all and were being steadily squeezed out of their jobs by economic forces quite beyond their control. But in general it could be said that the people were not averse to new ideas; some were even initiating change and discovering empirically whether or not it was economically sound. Also, the traditionally hospitable nature of the Omanis made them open and responsive to the influence of outsiders.

In a community development project the work has to be multi-faceted. The central importance of this had been stressed by Hutchinson (1966) who compared various schemes and found that the successful ones had in common a rise in living standards 'basically from the increased productivity of the community, largely in things they used themselves'. This of course requires a fuller use of local resources and the development of skills to process them. It also addresses the need for greater community independence based on mutual self-reliance. Additionally, it also allows the community to be aware of and to limit the 'disequilibria and observable lags in adjustments' that Schultz (1972a) had noted will occur under dynamic conditions of economic growth; growth in one sector will not necessarily have spin-off effects in others, so mutually dependent sectors each have to receive attention and appropriate support at the same time.

A multi-faceted approach was likely to have other more mundane advantages, important from the Project's viewpoint. First, it would allow more realistically for the fact that the Project's understanding of local potentials and problems would be imperfect, particularly at the outset. The possibility of failure, or relative failure, of some activities had to be anticipated when planning the Project. If a project is broad-based, such failures, hopefully in the less critical activities, can be accommodated, and this in turn helps relieve some of the pressure to get everything right first time. No project can achieve that ideal but if the central strategy is correct, the base-survey data sound, the interest and participation of the producers obtained, and the key activities going well, it can continue to carry conviction locally even if mistakes are made. Finally, a wider programme allows a project more readily to adjust its activities and its approach to changing circumstances. A project does not take place in isolation, and the social and economic environment in which it is operating will itself be under change due to other stimuli being applied in other parts of the wider rural system.

Fundamental to the approach was the acceptance that a 'proposed change must stand in some organic relationship to what the people are already doing' (Schumacher, 1973), working from the known to the unknown. Nearly all the enterprises of rural Oman were on a small scale: an owner-operator working by himself or herself, with his family, or with a very limited amount of paid help. It is these small enterprises which had to be reached if productivity was to rise. This approach is less spectacular and more difficult than the creation of a large new enterprise, and the achievement of results is slow. But as it involves a far greater proportion of the rural population, each working for himself within the village group, the results for the community and for growth appear more certain in the long run. Working at a small scale does not, of course, exclude an active interest in assisting small enterprises to grow or in helping large scale enterprises in appropriate ways.

Working from the known has other advantages. Most importantly it allows the producer to participate in the decision-making process because there are for him points of comparison with procedures with which he is already familiar. Radical departures are hazardous in a sector such as agriculture, where success depends on a complex mix of physical, economic and social factors. A one-step-ahead approach gives the opportunity of testing minimum changes in order to overcome significant bottlenecks, and to see what impact they have on the system as a whole before the next steps are attempted. Where the producer is part of the decision-making process he also retains his sense of responsibility for the consequences of changes that are undertaken and therefore a degree of personal obligation to strive for their success.

The central strategy of the project was to initiate interrelated schemes that encouraged the fuller but sustainable use of local resources by local people. This implied the development of the relevant skills and the growth of a sense of interdependence of those people involved. In turn it was hoped this would lead to a growing awareness that the village community, strong both economically and socially, was of central importance to Oman's future. In order to reach this point the village had to make a growing proportion of the things it needed. Additionally, it had to make money by selling produce outside the village in order to buy goods that were impossible to manufacture locally. Wherever possible, farmers had to be able to rely on village (or at any rate local) manufacturing units to provide their equipment. From a different perspective, craftspeople had to be encouraged to process local farm produce.

But how could the project meet the challenge encompassed by the development strategy given above? First, there was the foundation of understanding provided by the accumulated detailed knowledge of the region gained by the survey programme. It was also helpful that the first field manager of the development project, the author, had been a member of the survey team and therefore already had three years experience of the region. Again, in Oman the project had the broad sympathy of the Omani government and the logistic support of the oil company, Petroleum Development (Oman) Ltd (PDO), up to the end of 1985. Finally, the fact that project staff lived as well as worked within the village of al-Khabura and therefore were in continual contact with farmers and other villagers meant that they gained an understanding of people's individual abilities and interests and were able to respond to them rather than work to a theoretical developmental concept or to a programme determined by an evaluation of physical resources, such as land and water. As Schultz (1972b) had said, people, not land and water, are the key to development, or, in Schumacher's words (1973) 'people are the primary source of wealth'.

The project also had time. Both the Oman government and PDO wanted results, but neither made unrealistic demands for instant achievement. Time was essential if the aim was to reach a large number of people most of whom had no education and no experience of absorbing information in a formal learning situation. The people of al-Khabura were likely to view new ideas with appropriate caution and to accept them only as they became fully proven in practice. Apart from this, in any form of agriculture progress is partly governed by the seasons or by livestock cycles, and can only move at their pace. The main achievement of the project might then be not the acceptance of a particular technique but acceptance of the principle that change can be beneficial and that ideas stemming from the Project (or other sources) were at least worthy of serious consideration.

The approach thus outlined stands in contrast to but also complementary to a more remote and nationally planned approach to development. The latter has an essential place in all spheres of activity, but is most effective where dependence on the local physical environment and human element is smallest, as in the building of roads, ports, schools, etc. It is at its weakest where success depends upon people and upon the accurate understanding of many small-scale physical and human variables and their interrelationships - and also upon quick responses to changes in these - as in small-farmer enterprises and community development and the transformation of existing agricultural systems.

The project was located at al-Khabura for several reasons. al-Khabura was the single most fully studied village within the survey project; its soil and water resources, its agriculture, fishing and craft industries, its market and its demographic structure and change had all been the subject of detailed scrutiny. al-Khabura was at the centre of the Batina coast and since it was reasonably representative of conditions along the 270km coastal agricultural belt, replication, within that region at least, would have a good chance of success. The village also represented within its economy a wide range of indigenous activities each with its own problems, as the survey project had revealed, and therefore presented several potential choices of development programme. al-Khabura also had a more varied mix of people than any of the villages of the

Dhahira, and seemed to be more open to outside influences and accepting of change.

Agriculture almost inevitably was to be at the heart of the Project's economic activities in part because it involved, directly and indirectly, more people in al-Khabura than any other single sector. But in the 1970s agricultural work was too arduous and the rewards too small for the industry to hold the interest of the men in the most active age groups and they were migrating to the Gulf to find salaried work in ever greater numbers.

A major problem on the Batina, part cause and part consequence of the labour migration, was the falling market for the produce of the overwhelmingly dominant crop, date palms. Within Oman, as has been shown, people were seeking a much greater choice of foodstuffs, and so dates, particularly the poorer quality dates that grew on the Batina, were in much less demand. Many people now regarded Batina dates as fit only for livestock. In the market the better quality dates were trucked in from the interior of the country, or even imported from Basra. At the same time, the export market for dry dates had begun a decline in the early 1950s that had continued through the 1960s and into the 1970s. Because date production formed the basis of the only major farm system in northern Oman the reduction of the internal and external market for dates was very serious. Unfortunately the rapid fall in the value of dates, combined with the sudden increase in job opportunities outside the rural sector (and usually outside Oman), meant that there was neither the time, nor the obvious incentive, nor the necessary base of knowledge and experience to evolve new farm systems to suit new circumstances. Moreover, in addition to the market problems suffered by dates, the limes (a secondary but significant cash crop) suffered from increasing pest problems during the 1970s, particularly citrus blackfly and snowy-scale insects, with a peak in 1976/7 which decimated yields. Also, wheat, as has been shown, had become increasingly uneconomic to grow in the face of rising labour costs and cheap Australian imports.

Therefore, in agriculture the prime need was to devise and test new farm systems (FAO, 1995) to supplement or replace the one based on dates. It seemed appropriate therefore for the Khabura Project to turn its attention to this need, while remembering its other strategic priorities listed above.

Project resources were sufficient to concentrate on one small-farm system only, so it had to be one of central relevance to the needs of a large number of people. The Project had to provide not only a conceptual and practical bridge to the outside world but also a work base at and through which non-farm activities could be developed as well. It had also to possess sufficient logistical and working capability to be worth incorporating, at an appropriate future time, into the growing Omani agricultural, institutional framework.

The small-farm system that the Project decided to concentrate on was one based around the rearing and breeding of sheep and goats, and this for several reasons:

Within the agricultural sector, the activity which had the greatest number of 'upstream' and 'downstream' linkages was the one concerned with livestock. From methods of irrigation, cultivation and field layout to methods of forage management and harvesting, and from methods of feeding and managing the stock to using all their products including not only meat and milk but also skins, wool, hair and dung, everything could be scrutinised in the process of constructing a small-farm livestock system which the farmers would want and could, in practice, adopt. A livestock based farm system would also create opportunities to involve artisans manufacturing equipment needed by the farmers, and supply craftsmen with animal by-products to make produce for use in the village or for sale outside it.

Improved sheep and goat management potentially affected everyone in the village. Most households owned goats and sheep, two or three in the village, rather more in the gardens and up to fifty or more by the *shawawi* pastoralists in the gravel plain. All households, moreover, were obliged by tradition to slaughter a goat, or alternatively a sheep, at the *Id al-Fitr* and the *Id al-Adhha*.

The survey findings had shown that the traditional practice of having the animals scattered throughout the village was unsatisfactory in various ways. It was impossible for vets or other specialists in livestock husbandry to reach them. They carried dirt, flies and fly-borne diseases into the homes. They also fed on household scraps and other waste material, and although this reduced wastage it also led to unsuitable and even rotten food being eaten by the stock, causing intestinal diseases which were common.

At the same time, the other traditional practice of rough grazing and browsing sheep and goats, common in the plains and mountains of Oman, had also shown itself to be problematic. It had the merit of being cheap but also the major disadvantage, in a climate as dry and unpredictable as Oman's, of being totally dependent on the rainfall for the well-being of the stock. In a dry year goats and sheep could be so weakened by lack of food that they had no defence against diseases or predators, they produced no milk and they could not conceive or carry a foetus the full term. Worse still they endangered the sparse vegetation to maintain a carcass that had very little flesh on the bone.

The requirement was to change the attitude of stock owners from stock keeping to stock farming. This implied the creation of a farm system which incorporated a livestock unit of 50 to 150 head - large enough to give a significant income to a family enterprise and to justify veterinary and other services. Beneficially, rearing livestock in such a system meant that the well-being of the livestock would be less directly dependent on the natural environment. Management standards would have to improve so that productivity and quality would rise.

Since the relatively youthful official agricultural services were under enormous pressure and since the local scientific information base was still limited, project activities had to include applied research as well as practical demonstration, field testing of innovations and the monitoring and recording of a wide range of technical and human responses. The project operations had therefore to be based on a small farm, typical in its soils and water resources, and in size, location and terrain, of local farm holdings.

The Project's farm, however, was never intended to be a research station. In principle we acquired our local scientific knowledge of animal production from the MAF Livestock Research Station at Rumais, along the Batina highway 100km towards Muscat. In practice, however, the Project was not always able to obtain the information it needed from Rumais so it was pushed further in the direction of research than we wanted. For example, the Project had to make its own comparisons between local, exotic and cross-bred sheep and goats. We were not able at the outset to settle on animals of known quality under local conditions.

3.2 Elements of the farm and craft-industrial system

Activities were initially based on the Project's own farm. In practice it would have been unrealistic to ask the farmers and craftspeople to adopt the system until extensive field trials had, in continual discussion with the farmers and craftspeople, created something perceived by the farmers to be of real value, and until all elements of the system, and the system as a whole, had been properly demonstrated. Thereafter, we also hoped to work closely with interested members of the community on their own farms, or in their workshops.

The main elements of the farm system to be developed included, in addition to sheep and goats, forage crops and associated machinery, methods of cultivation and irrigation, design of farm buildings, locally made tools and equipment, veterinary care, dairying, spinning, weaving and leatherwork, and techniques for manuring and composting animal waste.

3.2.1 Farm purchase
But first, a suitable farm for the Project had to be found. After considering a number of alternatives, it was decided to attempt the purchase of a square of unused and unfenced land on the principal track that led from the heart of al-Khabura village to the new main road Figure 4). This seemed to have a number of advantages, including the fact that it was readily accessible from both the village and the road, and that everyone visiting the village market would pass alongside the farm and see, for better or worse, exactly what was taking place. The land lay in the midst of the main area of farm land around al-Khabura and demonstrably had soil and water of a quality comparable with those of its neighbours. Although bordered on three sides by date palm gardens, the land was, apart from a magnificent old *ghaf* tree (*Prosopis cineraria*) which remained a feature of the farm, entirely clear of vegetation. Therefore there would be no difficulty about constructing buildings or laying out the farm for livestock pens and forage fields. The land, under the then ownership of one man, was of an

appropriate size, about 2.2ha, to allow for staff housing and farm buildings and still leave an area for production that lay within the normal range for the region. Accordingly, the owner was approached (he was a man long known to us as a merchant in the *suq* and the owner of nearby palm gardens) and although he drove a hard bargain the land was successfully purchased, and the Project had full freedom to develop the land as it saw fit.

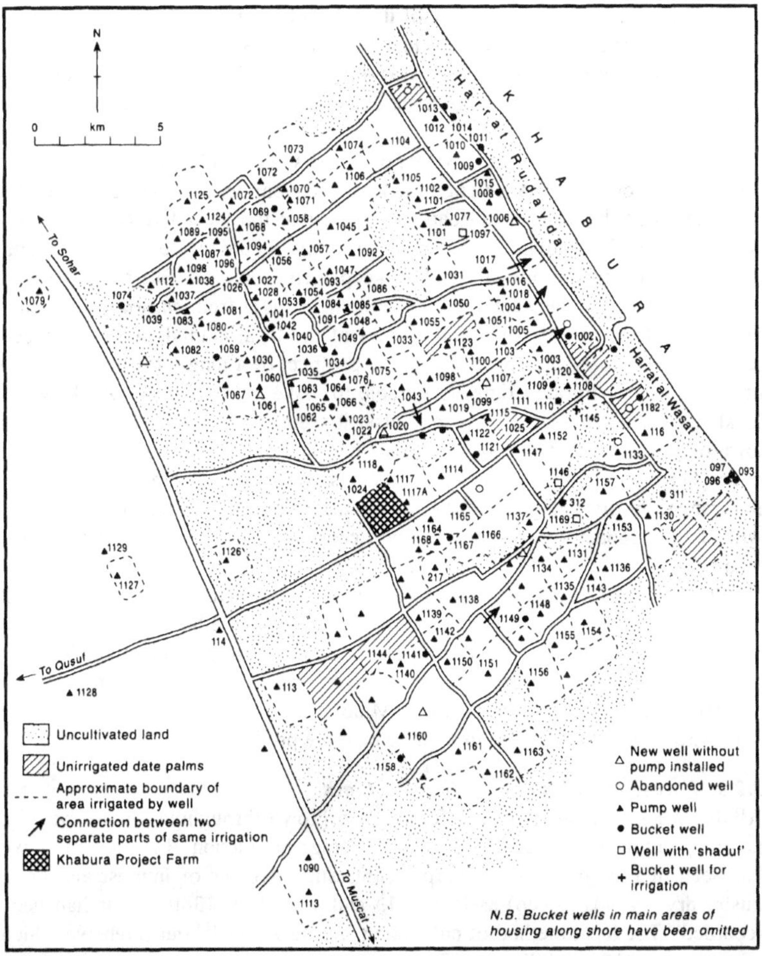

Source: Letts, 1978b

Figure 4. Small farms near al-Khabura, 1973/4, showing wells and the location of the future Khabura Project's farm

3.2.2 Goats and sheep

Because the survey project had shown that in the al-Khabura region of the Batina coast there were approximately four sheep to every six goats it seemed appropriate to work with both animals. Goats are traditionally regarded as the principal small ruminant in Oman, but those few farmers who kept significant numbers of small ruminants had commented that in farm conditions sheep had certain management advantages over goats, being easier to pen and less susceptible to disease. The survey of the *Id* markets in 1974/5 had also shown that for the range of medium-sized female goats and sheep (the majority of the animals that were on sale) there was little discernible price differentiation between the two. Only the large male goats had commanded a premium price (Dutton, 1982f).

The main aim was to build a system of sound and sensible, more intensive but not over-elaborate, management practices (Economides, 1983) based on the basic needs of the sheep and goats, a system that the local farmers could and would eventually emulate. The main emphasis would be to feed sufficient, nutritious food to the animals and to keep them healthy. But a comparison would also be made between the local animals, bought from *shawawi* and small farmers in the central Batina, and crosses of the local animals with one breed each of exotic goats and sheep. With the prospect of a long-term project ahead of us we hoped, through time, to be able to make valid comparisons between local stock and cross-breds. At the outset it seemed likely that the adaptations towards hardiness that the local animals had had to make in order to survive under Omani minimal management systems, where the sheep and goats had free-ranged in the villages and the mountains, might have produced an animal with limited capacity to respond to good conditions under semi-intensive farm management.

We accepted a recommendation to import the Anglo-Nubian goat as a general purpose breed because it was thought to be tolerant of the hot, humid and dusty conditions found in Oman yet to be significantly more productive than the Omani goats. Two breeds of sheep were given serious consideration, the Awassi and the Chios. The Awassi, a major Middle Eastern breed, was known to be very hardy but the Chios was thought to be more productive.

3.2.3 Forage crops and harvesting machines

Alfalfa, a leguminous perennial forage crop grown throughout Oman, was the natural point of departure for the Project's forage production. In al-Khabura, we had seen, it was grown with sharply contrasting degrees of interest and good husbandry - yields (green) as low as 15 and as high as 134t/ha/year had been recorded. The local method of cultivation was also very labour intensive which put its continuing viability at risk at a time when labour was becoming more scarce and expensive, and family labour increasingly reluctant as children became more concerned with their school work than with helping their fathers on the farm. The peak summer temperatures were also a little too high for alfalfa. Any soil water stress resulted in wilting and checked growth rates which gave the opportunity for tropical weed grasses to invade and dominate the crop. So there was a need to raise mean yields by improving the average standard of husbandry, and a need to see whether the labour input could be reduced by the use of appropriate machinery for harvesting the crop.

We decided also to introduce to the Project (and, incidentally, to Oman), Rhodes grass as an alternative to alfalfa, on the basis of the findings of research and development work undertaken at Hofuf in Saudi Arabia. Rhodes grass has the disadvantage that it is not a legume and therefore, in the nitrogen poor soils of Oman it would require the application of nitrogen fertilisers (Farnworth and Ruxton, 1974), an extra complication and one with which the farmers were totally unfamiliar. It is also not as nutritious as alfalfa, but for feeding to sheep and goats (as opposed to cattle) this was probably not a disadvantage. Rhodes grass also has important compensating advantages. First, it is very dense in its growth and was likely therefore to smother any grass weeds trying to compete with it. Second, like alfalfa, it is perennial and therefore would not require the cost and labour of annual reseeding. Third, it has a vigorous system of vegetative reproduction by stolons which makes Rhodes grass an ideal reclamation crop. Rhodes will colonise bare patches of land resulting from initial salinity, as successive irrigations leach excess salts from the topsoil, but not become an almost irremovable weed as happens with rhizomatous grasses like Bermuda grass. Rhodes grass is also tolerant of a high level of soluble salts in the soils and irrigation water, which are characteristic conditions on Oman's Batina coast. Furthermore, it is very tolerant of drought and so suited to gardens where poorish management and occasional pump breakdowns are likely to result in a somewhat irregular irrigation regime. Rhodes grass positively thrives in the summer heat of Arabia, and any surplus summer production obtained at al-Khabura could easily be turned into hay by leaving it to dry for a day or two in the sun. Rhodes grass was also likely to be suited to grazing by both sheep and goats, a feature which would reduce requirements for labour and machinery, allow the animals to feed on demand, and return dung and urine to the field.

To harvest both alfalfa and Rhodes grass, we judged that a pedestrian-controlled reciprocating-blade forage cutter would be a suitable piece of machinery to test and demonstrate as a means of cutting the crops, to see whether it would attract the interest of the local farmers.

Supplementing Rhodes grass with alfalfa in livestock feed during late pregnancy and lactation was anticipated. Alternatively the Project would have the option of dates as a locally grown concentrate, or of using processed concentrates newly being produced by the Oman Flour Mills.

3.2.4 Cultivation and irrigation

At the time the Project started there were two local options for land cultivation. We could either use one of the bull ploughs to break and turn the earth followed by hand preparation of the seed bed, or hire the extension centre four-wheel tractor to turn the soil, again followed by hand preparation of the seed bed. For the inaugural ploughing, the Project envisaged hiring the tractor, partly because the land had not previously been cultivated and might benefit from being broken at depth by the extension centre deep tyne which only the tractor could pull through the soil.

On the principal of working from the known to the unknown the project also decided to use the traditional method of distributing irrigation water - from the well along water channels to irrigation basins. However, for the reasons suggested by the small-farm survey, the Project decided to increase the size of the basins to 20m by 10m in the hope that they would save land, water and time, facilitate the use of machinery and significantly increase total production. It was recognised that additional savings of land and time would most readily be made by changing from irrigation basins to border strips, or eventually to piped irrigation systems. But the former would have created problems of unequal water distribution within the basins, and the latter, involved, in 1976/7, too great a departure from the known to be given serious consideration. Piped irrigation would have required the establishment of a supply and financing system, the provision of installation, maintenance and repair skills, and careful consideration of the likely consequences for soil-plant-water relations.

As the decision had also been made to use a modified form of the traditional flood irrigation (Kay, 1986) it was accepted that shaping the *gelbas* and the irrigation channels would require a lot of manual labour, particularly when they were being made for the first time. But, it seemed, there was also a niche for a pedestrian-controlled two-wheel tractor, or rotavator, significantly to reduce the labour requirement and prepare a better tilth. Such a machine would reduce in one pass the clods left by the plough to a good seed bed. Also future re-cultivation could be done entirely be the rotavator. Its use would encourage large basins to be made, because the rotavator would be able to work within a basin (unlike the tractor) and cause little or no damage to neighbouring plots, irrigation channels and basin walls. Larger basins, as the survey project had shown, would be much less wasteful of land and water than small basins. As the rotavator would cause them minimal damage there would be a much greater incentive to recultivate and recrop them more frequently - as we had demonstrated, one of the best and simplest ways to increase average yields of perennial crops in Oman (notably alfalfa) would be to resow old stands before they became weak and unproductive.

3.2.5 Farm buildings
We discovered during the survey project that the livestock housing, typically, was poor or non-existent. Normally in the villages no special penning or housing provision was made for the goats or sheep, other than perhaps a rudimentary bit of shade within the domestic house compound. The village cows fared rather better though often they were tethered in too confined a space, and their stalls were not mucked out sufficiently often to keep a check on

parasites. In the gardens the housing was equally rudimentary, though in some cases it was being replaced with airless heat-retaining concrete block structures. There was, therefore, a real need to test the use of locally available building materials and experiment with designs and layouts of improved livestock housing. Pens were needed to suit the animals and simplify the tasks of feeding, watering, mucking out, and moving the animals between pens and other facilities.

3.2.6 Locally made tools and equipment
The survey had shown that tools and equipment, in so far as they had been used by local farmers, had mostly been manufactured by local craftsmen, such as basket and rope makers, blacksmiths, leather workers, and ploughmen. But by the mid-1970s most of these craft professions had fallen into abeyance and with this decline had gone the mutual dependencies which previously had linked farmer and craftsman and the materials on which the craftsman depended. Instead, farmers were buying equivalent tools which were mass-produced, cheap and imported. But the new small-farm livestock system that the Project hoped to test and demonstrate provided the opportunity to design a new range of tools and equipment for local manufacture in the new workshops and garages that were appearing. The outcome would serve the farmers needs, create local employment and rebuild some of the mutual reliances between skilled people that had characterised the original rural community.

3.2.7 Veterinary
The survey project had shown that the almost complete absence of veterinary support for the livestock in the vicinity of al-Khabura had created a situation in which, at the worst in a given village, over half the herd could die in any one year. This was not only a very important loss of wealth but also a powerful disincentive to making serious investment in even small-scale livestock enterprises. Large losses were tolerable only where the sheep and goats were fed largely on scraps and grasses or left to fend for themselves with few if any overt costs involved. But losses had to be minimised if farmers were to pay the relatively high costs involved in building and maintaining a good quality and productive herd of, say, 50 head or more.

Therefore veterinary care had to be part of the small-farm system under examination. The provision of this service offered an opportunity for the Project's staff to make dirèct and practical contact with a large number of livestock owners and to train one or more local people in basic veterinary skills.

3.2.8 Dairying
We had seen that nobody kept either sheep or goats for their milk. The small ruminants were regarded as meat animals, and the cow as the supplier of milk. Nevertheless, when the winter rains were good and had created plentiful grazing and browse so that the *shawawi* goats were in good condition and producing

milk surplus to the needs of their kids, the *shawawi* would take a little milk from the goats and either drink it straight or in tea. We knew therefore that goats' milk was liked, and we also knew that the well-fed goats on the project farm would produce a higher milk yield than the goats in the villages or on the plain. This created the possibility of a small-scale, perhaps village based, dairying enterprise, and the opportunity to create a new mutual dependence between local producer and consumer.

3.2.9 Spinning, weaving and leather work

In contrast with dairying, spinning and weaving had been widely practised locally. The wool had been an additional source of income for those people owning sheep. The women of the *shawawi* households, using hand-held spindles, span and plied the yarn, and the male weaver in Rudayda wove the traditional local rug or *mansul* for them. It was a closed, local economic activity, and little if any attempt was made to sell either the *mansuls* or the yarn to people from further afield. But although the weaver stopped work for good in 1976, and although only a few women still span, many people in all the *shawawi* settlements retained their inherited spinning and weaving skills. This offered an opportunity to revitalise the local industry and by some combination of reducing the labour requirement, increasing productivity, improving quality, extending the range of goods, and seeking new markets, see whether spinning and weaving had a future role. If so, it would, once again, turn wool into a valued by-product and create a community of mutual dependence between primary producer and craftsperson. In this case the work would recreate an economic activity for women, one from which they would earn money directly and have control over its usage.

Leather work, like spinning and weaving, was shown by the survey to be widely practised in northern Oman. Both men and women of the *shawawi* families knew how to tan animal skins and to make a small range of products valuable to the household, notably the milk churn. In addition, professional leather workers made the big leather *dellu* for lifting water from the wells for irrigation. The opportunity therefore existed for trying to reactivate the dying craft by, perhaps, introducing new technologies or making new products to meet the needs of new markets. At the same time this would, of course, put extra value on the animal and perform the community function of linking primary producer with craft processing local industry.

3.2.10 Manure and composting

A benefit of keeping larger numbers of animals in properly designed pens is that their waste matter and any litter from uneaten forage can easily be collected together, removed from the pens, converted into compost and reapplied to the field. This ability to recycle biological waste would be an important part of the system under trial.

4. The Khabura Project: field trials, demonstrations and training

4.1 Goats and sheep

The purchase, breeding, management and value of the Project's goats and sheep are here discussed, including cross-breeding programmes with local farmers' goats.

4.1.1 Purchase

Having made the decision to import Anglo-Nubians, the Project needed to ensure good quality stock. In late 1976 we approached Mr Derek Kibble, a great enthusiast for the Anglo-Nubian breed of goats in the UK. He helped us select an initial batch of two males and three females with good kidding and milk records, and export them to Oman[1]. On January 15th 1977, a very memorable day, they reached the project farm by truck from the airport and were the first animals to be accommodated by the Project, in new pens that had hastily, but carefully, been constructed. In 1982 we imported an additional seven Anglo-Nubians, three pregnant females and four males of different blood lines[2].

Up until 1982 the Project had primarily regarded its goats and sheep as meat animals, which is how they were normally regarded in Oman, but from mid-1982 we began to explore their potential as producers of milk for sale. Thus a specialist was appointed in 1983 for this work (Ch. 4.3) and, in 1985, consideration was given to importing a British Saanen male to breed with the Anglo-Nubian and local crosses in order to increase milk yield. Experience in India (letter from Mrs Patricia Sawyer, 1985) suggested that the results would be good, but in practice we did not follow through.

As a direct result of the discussions held with MAF in the months prior to the first Anglo-Nubian imports, MAF itself imported Anglo-Nubians through the same supplier in 1977 and again in 1979[3]. By exchanging goats with MAF we were then able to increase our own range of blood lines.

---

[1] Derek Kibble was the then owner of the Marathon Herd of pedigree dairy goats, in Coggeshall, Essex. The goats cost £575 in total, while transport including crates cost £893. The goats were Nenevalley Fanta and Bugle Baron (male), and Glenside Glengavin, Glenside Glengairn and Alderkarr Excella (female). In their 1976 kidding Excella produced triplets and over 2,000lb milk, Glengavin twins and 8lb milk/day, and Glengairn twins and also over 8lb milk/day. In practice, only Excella was pregnant on arrival in Oman.
[2] Selected for us by Mrs D Sawyer, Castle Cary, Somerset.
[3] MAF imported 10 males and 10 females through Derek Kibble later in 1977, and a further 20 females in 1979. With these animals they undertook on-station research at their main livestock research station at Rumais, southern Batina.

Concerning exotic sheep, the Project made an early decision to import Awassi sheep because they are a general purpose Middle East breed known to be very tolerant of harsh living conditions. However, in spite of making approaches in various Middle East countries, and as far afield as Cyprus and Spain, we were not able to find animals of good quality and confirmed disease free, and/or at an acceptable price, thus by mid-1979 thoughts were turning to other breeds. By then it had become clear from the Project's experience of goat rearing and sales that in the then structure of the Omani livestock market it was far more profitable to have a large number of smaller animals rather than a small number of larger animals. This made the Awassi seem less attractive because it was too large and not sufficiently prolific, and we were also becoming aware from trials conducted by MAF that the Awassi was not popular with the Omani farmers because they had no familiarity with handling such a large animal and because they did not know what to make of the Awassi's fat tail. We turned to many people for advice, both inside Oman and internationally, including Mr Ian Mason[4] who suggested on the grounds of prolificacy, wool type and regional origin either the D'man of Morocco, the Gallego of Spain or the Chios of Greece. For the latter he gave the adult weight of the ewes as 48-52kg, prolificacy as 180%, wool of 8cm staple, and with a high milk yield. They seemed therefore perfectly to suit the Project's requirements though no one had experience of them in the very hot and humid conditions of coastal northern Oman. FAO advised us to proceed with caution, but expected no adverse effects on their performance from the climatic conditions[5]. We therefore decided to go ahead, and placed an order in November 1980 with the Agricultural Research Institute in Cyprus for: 'five Chios sheep including two half grown ram lambs and three ewes of prolific line that are in lamb from males other than the two we are purchasing'. Transport and other problems delayed shipment for another full year, but we finally took delivery of the animals as ordered[6], and in February 1982 they produced a singleton and two sets of twins which were fed a non-special diet and put on weight up to end-May at 300gm/day. Neither the young nor the adults were showing any undue heat stress up to that time. The first cross-bred lambs were born later that year. The Chios and the cross-breds continued to thrive. In the Project's first appraisal period in mid-1983 we noted

---

[4] Ian Mason, previously a consultant on animal genetic resources at FAO, had become a freelance animal breeding consultant, based in Edinburgh.

[5] In April 1980 A.S. Demiruren, Animal Production Officer at FAO, wrote, 'Concerning your query about the suitability of Chios for introduction into Oman; we have not yet tried them under such ecological conditions. However, if they are looked after well and their nutritional requirements are met I expect no adverse effects in their performance. The first shipment should be for testing their adaptation and the importation of larger numbers should follow the successful adaptation of the breed'.

[6] Rams: (1) born as a triplet, dam had completed four lactations and given nine lambs and 1,360kg of milk; (2) born as a twin, dam had given in two lactations four lambs and 790kg of milk. Ewes: born in 1978 and had given in two lactations, respectively, 4 lambs (840kg milk), 3 lambs (450kg) and 4 lambs (500kg).

that the Chios had settled in very well and coped with the rigours of the summer climate with surprising ease. One or two imported males had died with lungs congested with pneumonia but otherwise no imported adults or pure or cross-bred lambs had perished. The cross-bred lambs were bigger at birth - without causing any birthing problems - and visibly superior in growth rate and size to the Omani lambs. Reports to the end of 1983 were sent to various of the advisers who had helped in their selection. Professor Goot[7] commented: 'Your results in Oman are surprisingly good ... you will probably run into health and thrift problems'. John Maule[8] commented: 'There may also be a place for the pure-bred Chios ewe, which seems to have acclimatised well to Omani conditions. Perhaps you will be able to make a comparison between the two pure breeds [Chios and local] as well as between the cross-breds'. Ian Mason commented: '...I am glad that so far cross-breeding is proving successful ... the production of 3/4 Omanis does not seem necessary from the practical point of view, in view of the good performance of the 1/2 and pure Chios'. Mr Economides[9] commented: 'I was delighted to read that Chios sheep survived under the high temperature and humidity in Oman ...The rate of growth of Chios lambs was excellent until 3 months of age, but thereafter was reduced particularly in males between 3-6 months'. Oman's MAF also decided to purchase Chios sheep, but never in fact did so. Therefore there was no possibility of exchanging animals with them.

Purchase of local goats and sheep started during the spring and summer of 1977 and continued, sporadically, throughout the life of the Project. By the end of 1978 the number of exotic, local and cross-bred stock had reached 60. The total exceeded 100 by the end of 1979 when the estimated stock capacity of the farm's 1.7ha of forage crops was tentatively revised upwards to 200. In fact the total number of goats rose to a maximum of 170 in 1988, and the number of sheep to 160 in 1986 - some 300 animals all told during the late 1980s though by then forage was also being locally purchased, and concentrates formed a significant part of sheep and goat nutrition.

4.1.2 Breeding systems and management
Initially, not wishing to move too far from the breeding system with which the local farmers and pastoralists were familiar, the Project adopted a policy of natural breeding for its sheep and goats. But as the stock numbers increased, the impracticalities of natural breeding became manifest. Oman enjoys a relatively even day length summer and winter and so there is no natural breeding season. Goats and sheep are capable of conceiving at any time of year. In consequence lambs and kids were born in all months, and the young and the adults were all at different stages of their growth or breeding cycles, and therefore needing

---

[7] Rhe Volcani Center, Israel.
[8] Livestock information consultant, Scotland.
[9] Agricultural Research Institute, Cyprus.

different diets (Steele, 1983a). Management to optimise feeding regimes was impossible. Therefore we decided to explore the option of controlled batch breeding both goats and sheep (Gordon, 1983) using hormonal control. We hoped that the flock would all breed on or near the same day and that the adult females could be prepared for remating within three months of giving birth, thus achieving an eight month breeding cycle - or three lambings and kiddings in two years. In this way the Project would turn the lack of breeding season to advantage.

Table 1. Value of kid crop and efficiency of production[1]

Production parameter	Pure local	X-bred (local sire x X-bred dam)	X-bred (Anglo-Nubian sire x local dam)
Mean litter size	1.26	1.76	1.26
% kid survival to one year	81.7	80.4	79.9
Number kids for sale at one year	1.03	1.42	1.01
Mean kid weight at one year (kg)	28.6	32.5	36.1
Total weight of kids sold (kg)	29.4	46.0	36.3
Example price/kg liveweight (RO)	1.751	1.519	1.519
Total output value (RO)	51.552	69.857	55.206
Dam's mature weight (kg)	38.2	46.8	38.2
Dam's metabolic weight[2]	15.37	17.89	15.37
Efficiency of production[3]	1.92	2.57	2.37

Source: Taylor, 1990a

(RO per kidding female, assuming a 365 day kidding interval)

(1)   Assumes 50% offspring male and 50% female, and all surviving kids sold at 365 days
(2)   Body weight to the power 0.75, and closely related to the maintenance feed requirement
(3)   Weight of surviving offspring at 365 days as % of dam's mature (1095 day) metabolic weight

The systems tested included injections with Estrumate and the use of hormone impregnated vaginal sponges, with and without an injection of PMSG[10] to increase litter size. The first trials for both sheep and goats, initiated in 1982 and 1983 (Steele, 1983a-c), were sufficiently encouraging - from the points of view of productivity and batching - for a 5-year trial, 1983-8, to be undertaken, principally under the mangement of Harvey and Allison Sherwood (Sherwood, 1984). The double objective of the long-term trial was to monitor the impact on herd management and on reproductive performance. We deliberately took the opportunity to test the most demanding system, three lamb/kid crops in two years. Realising that if a ewe or she-goat failed to conceive it would lose eight months before it could be mated again with the whole herd, we divided the total herd into two (flocks A and B), with breeding cycles staggered at 4-month intervals. Thus if an animal failed to conceive it could join the other herd and lose only four months. Two successive breeding failures resulted in the animal being culled. Having a double herd also had the advantage of halving the intensity of the workload at each stage of the cycle - and in particular it meant

---

[10] Pregnant Mare Serum Gonadotrophin

that more care could be given to the animals during birthing. In practice the staff operating the system also found that because of batching it was easier to perform a variety of other operations as well, such as vaccinations, ear tagging, etc.

During the period 1977 to 1983, the goats and sheep had 'free' bred at almost any time of the year, usually as they returned to oestrus after weaning. From 1983 mating was organised with the intention that lambs and kids would be born in June, October and February (avoiding the combination of heat and humidity that characterised July to September). To facilitate batch conception about 30% of goat matings and 64% of sheep matings were synchronised with Chronogest vaginal sponges. Additionally 40% and 57% of local goats and sheep were injected with PMSG.

Table 2. Value of lamb crop and efficiency of production[1]

Production parameter	Pure local	X-bred (local sire x X-bred dam)	X-bred (Chios sire x local dam)
Mean litter size	1.20	1.20	1.39
% lamb survival to one year	84.7	85.7	89.7
Number lambs for sale at one year	1.02	1.03	1.25
Mean lamb weight at one year (kg)	32.3	42.8	46.4
Total weight of lambs sold (kg)	32.8	44.0	57.8
Example price/kg liveweight (RO)	1.431	1.179	1.179
Total output value (RO)	46.979	51.894	68.209
Dam's mature weight (kg)	33.0	33.0	45.4
Dam's metabolic weight[2]	13.77	13.77	17.49
Efficiency of production[3]	2.38	3.20	3.31

Source: Taylor, 1990b

(RO per lambing female, assuming a 365 day lambing interval)

(1) Assumes 50% offspring male and 50% female, and all surviving lambs sold at 365 days

(2) Body weight to the power 0.75, and closely related to the maintenance feed requirement

(3) Weight of surviving offspring at 365 days as % of dam's mature (1095 day) weight

One great advantage of operating a long-term project was that we were able to analyse sheep and goat production and sales data for the whole period 1977-89. A total of 927 goats and 843 sheep were born on the Project's farm, divided between exotic, local and cross-bred. Records were well maintained throughout the period, and thus we had a good statistical basis for comparing the value of the different breeds and types and comparing the two breeding systems. Using the actual values for litter sizes, survival and growth performance achieved for goats and sheep at the Project's farm, and the actual prices received for kids and lambs in the local market, output values and production efficiencies were calculated for goats and sheep as illustrated in Tables 1 and 2.

Table 1 shows that the Anglo-Nubian genes made an important contribution to weight of kids sold, total output value and efficiency of production. The best financial result was obtained from local males on cross-bred females. This yielded a 35% increase in sales value over the local kids and a 34% increase in efficiency of production. Table 2 shows that the Chios genes also made a very important contribution to weight of lambs sold, total output value and efficiency of production. The best financial result was obtained from cross-bred males on cross-bred females. This yielded a 45% increase in the sales value over the local lambs and a 39% increase in efficiency of production.

Table 3.   Yearly value of kid and lamb crop, and efficiency of production[1]
(RO per kidding/lambing female, at different birthing intervals)

	Yearly	9 months	8 months
Local goats			
Number kids for sale at one year	1.03	1.37	1.55
Total weight of kids sold (kg)	29.4	39.2	44.2
Example price/kg liveweight (RO)	1.751	1.75	1.75
Total yearly output value (RO)	51.552	68.7	77.4
Efficiency of production[2]	1.92	2.55	2.88
Local sheep			
Number lambs for sale at one year	1.02	1.35	1.53
Total weight of lambs sold (kg)	32.8	43.7	49.3
Example price/kg liveweight (RO)	1.431	1.43	1.43
Total yearly output value (RO)	46.979	62.58	70.57
Efficiency of production[2]	2.38	3.18	3.58
Cross-bred goats			
(local sire on cross-bred)			
Number kids for sale at one year	1.42	1.89	2.13
Total weight of kids sold (kg)	46.0	61.2	69.0
Example price/kg liveweight (RO)	1.52	1.52	1.52
Total yearly output value (RO)	69.9	93.0	104.9
Efficiency of production[2]	2.57	3.42	3.86
Cross-bred sheep			
(cross-bred sire on cross-bred)			
Number lambs for sale at one year	1.39	1.85	2.09
Total weight of lambs sold (kg)	57.8	76.9	86.7
Example price/kg liveweight (RO)	1.18	1.18	1.18
Total yearly output value (RO)	68.2	90.7	102.3
Efficiency of production[2]	3.31	4.40	4.97

Source: data from Taylor, 1991a

(1)   Assumes 50% offspring male and 50% female, and all surviving lambs sold at 365 days
(2)   Weight of surviving offspring at 365 days as % of dam's mature (1095 day) weight

The increases in sales value, for both kids and lambs, were achieved in spite of the fact that the cross-bred kids and lambs were bought (for slaughter) at a significantly lower price per kilogram than the pure, local kids and lambs.

Table 3 illustrates the production and financial gain if the goats and sheep are actually bred at 9-month or 8-month intervals, instead of annually. In practice, during the 5-year frequent breeding trial the median lambing interval for local and cross-bred sheep was about 245 days. This shows that under the practice of synchronised frequent breeding it was possible to maintain an 8-9 month breeding regime. Some 76% of lambing intervals were less than nine months though some 24% clustered around 12 months. However, comparably short lambing intervals were also possible under 'random' breeding management with 67% of intervals being less than nine months, though this was without the benefits of batch lambing, outlined below. The technique, therefore, did not result in much extra output in terms of lambs per year and, in fact, also led to more breeding intervals longer than 12 months (due to the necessity to batch breed). For goats, on the other hand, synchronised frequent breeding did reduce kidding intervals, compared with 'random' all year round breeding, but less than 30% of kiddings fell within the targeted 8 months. Although over 50% were on target for 9-month kiddings, making a 9-month kidding interval a feasible aim, it was concluded that annual breeding might be more productive long-term (Taylor, 1990a).

But for both sheep and goats the practice of running two parallel herds prevented kidding and lambing seasons drifting. Most importantly, remarkably tightly batched kidding and lambing patterns were achieved. Kids born during and after 1983 had higher weights at all ages than kids born under the free breeding regime prior to 1983. As Taylor (1990a) states: 'This change is quite striking' and is most probably associated with the batch breeding which permitted much better stock management, especially control over nutrition. A similar improvement in lamb growth performance was almost certainly due to the same set of management changes which batch breeding permitted. Taylor (1991e) also points out the importance of having a mating period lasting at least a full two oestrus cycles (the mean on the Project's farm was only 1.5).

Taylor's overall conclusion was that for a small farm unit (fewer than 50 breeding females) the management advantage of batch breeding would be minimal - random mating August to December would maximise the pregnancy rate without causing too many management problems. However, larger commercial sheep herds would give best breeding results if tightly batched in two parallel groups on an eight month schedule, though the results also showed that a goat management system based on an eight month kidding schedule would not be sustainable. Very large herds in which the need for streamlined management probably outweighs the need to maximise production should probably go for tight batching based on a 12-month lambing/kidding interval.

Nutrition was always a major concern (CAB, 1980). As might be expected, over the life of the Project, the goats and sheep were given a variety of diets. However, it was not possible to control the feeding regimes in the years up to 1982 when kids and lambs were being born throughout the year. Improvements in housing did allow differential access to feed, but it was only after the introduction of the batch breeding programme in 1982/3 that it became practically possible to adjust the feed regimes to suit the nutritional requirements of growing lambs and kids and mothers at different stages of their reproductive cycles. The basis of the diet was Rhodes grass together with some alfalfa and small amounts of other tropical grasses and legumes. Mostly the forages were machine cut and stall fed, though grazing was an increasingly used alternative.

Also, surplus green forage produced during the peak yielding summer months was dried to form hay and fed to the livestock in the subsequent winter when forage production was at its lowest. Rhodes was fed more or less *ad lib.*, with adults receiving around 2kg freshweight per day. In addition, concentrate feed was purchased from Oman Flour Mills (general ruminant feed; approximately 15% protein). This was provided on an approximate bodyweight maintenance and production basis; a 50kg non-pregnant adult being given about 0.25kg per day for maintenance, and pregnant females up to 1kg/day. Locally available dried fish and dates were also fed, depending on availability and the interest of the farm manager at any given date. From the early 1980s creep feeds were installed to initiate kids and lambs on concentrates while still suckling.

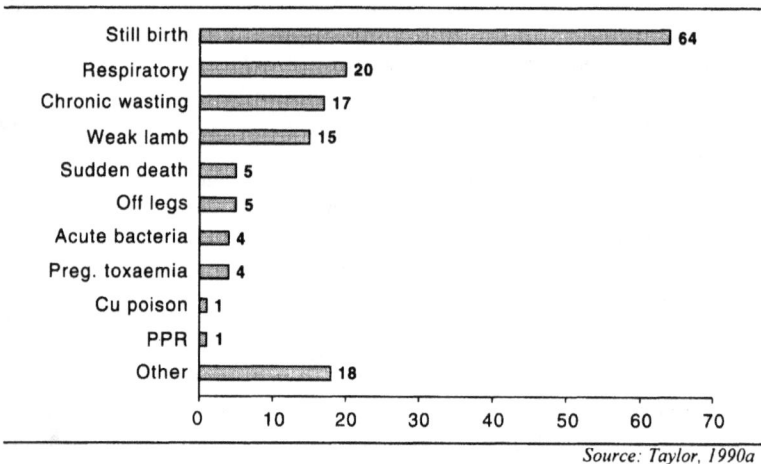

Source: Taylor, 1990a

Figure 5. Deaths by different causes in sheep on the project farm, 1977-89

Diseases and deaths were carefully monitored. Goat and sheep 'departure' cards from the project farm recorded, as well as sale and exchange information, data on all deaths which occurred, including at least a tentative diagnosis of cause of death. Death rates were then calculated to express the number of deaths as a proportion of animals 'at risk'[11] within a set of specified age groups which included at birth, 1-90 days, 91-365 days, 366-730 days and older groups (Taylor, 1990a-b).

As can be seen in Figure 5, the main four causes of deaths in sheep were:

- still births (or 'at birth') - affecting local, cross-breds and Chios;

---

[11] Animals at risk were calculated via the following steps: (a) total the number of days animals of a particular type were present and within the age group. (b) divide by length in days of the age group e.g. by 90 days for the 1-90 day age group.

- respiratory disease - mainly in Chios and crosses, and in the pre-weaning age group;
- chronic wasting, or ill-thrift in older sheep - mostly local sheep over three years of age, (may have been due to Johne's disease);
- weak lambs - affected all types, (possibly causally related to still births).

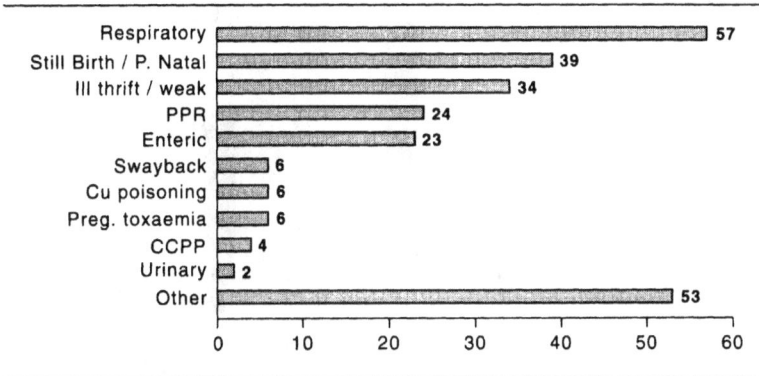

Source: Taylor, 1990b

Figure 6. Deaths by different causes in goats at the project farm, 1977-89

In goats, Figure 6, the main causes of death were:

- respiratory disease, chiefly pneumonias and pasteurellosis (Hall, 1982) - mostly between weaning and one year, and tending to be concentrated in the winter/spring months in association with wet weather;
- still births and peri-natal deaths - almost all cross-bred and Anglo-Nubian kids;
- ill-thrift - mostly in suckling kids in the winter months;
- Peste de petits ruminants (PPR) - 24 deaths in one outbreak in April/May 1987, almost all young cross-breds and Anglo-Nubians; and,
- enteric disease, mainly bacterial scours (a year-round problem) - all cross-breds and Anglo-Nubians.

This pattern of disease is typical of that found in housed ruminants anywhere in the world but it is notable that although stillbirths were rather more common in sheep than goats (Figure 7) deaths of sheep were markedly less frequent than of goats in all other phases of their growth to maturity. Within the sheep, an examination of the different breeds and crosses revealed that Chios lambs exhibited a rather higher death rate but that the cross-breds once weaned survived slightly better than the locals (Taylor, 1990b). Above two years of age the local sheep exhibited a high death rate, possibly linked to the presence of Johne's disease on the farm. For the goats (Taylor, 1990a) losses of pure Anglo-Nubians up to two years of age were high. However, there was no meaningful trend separating death rates in local goats from the different crosses. The net result was that while 70-80% of cross-bred goats survived to a year old, a significantly higher 85-90% of the cross-bred sheep survived to that age.

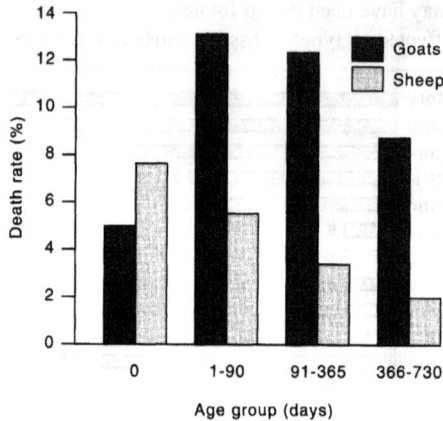

Source: Taylor, 1991a

Figure 7. Overall death rates (DR%) in different age groups of sheep and goats on the project farm (Birth = stillbirths as % of all births)

There was a very important 'year effect' in sheep and goat death rates. In the period from 1982, pre-weaning death rates were mostly lower than before. Even more interestingly, in the years from 1982/3 the post-weaning to 365-day death rates fell significantly for sheep and very dramatically for goats. Therefore, during the time when the farm was running at maximum capacity, death rates were generally within reasonable limits. This probably reflects an improvement in overall management, including disease management, due to batch breeding and an overall streamlining of the farm system.

Lambing and kidding problems are another consideration when choosing between breeds and types of animal. For each kidding and lambing the project staff recorded whether or not assistance had been given. No incidence of assistance for kidding in any group rose above about 10%. For sheep, however, Chios sires on local ewes led to 25% of lambings being assisted. The rate of cross-bred on cross-bred assisted lambings was also high, at 12%. These high figures are probably also associated with the relatively large proportion of stillbirth lambs (see Figure 7), and would be a significant management concern if the Chios (or other large sheep breeds) were to be used more widely for cross-breeding in Oman.

4.1.3 Goat and sheep sale prices, and values

In the economic analysis given in a later section of this chapter (Ch. 4.12) the sale prices in the model are the mean prices achieved by the project farm for sales of local, exotic and cross-bred goats and sheep. Local goats for slaughter achieved the highest price for any group of animals sold, at RO1.75/kg (Table 4). When sold for slaughter the price per kilogram of both local goats and sheep was more than for the cross-bred equivalent. However, the Project's sale prices probably underestimated true local sale prices because the staff did not have the time or local experience to haggle. Also, the prices probably exaggerate the differential between the buyers' evaluation of local as opposed to cross-bred animals because many of the sales were made at the time of one or other of the main religious festivals when tradition counted most strongly. The sale price for cross-breds for breeding was markedly higher than for local stock which indicated the value that the farmers placed on the greater productive capacity of the crossed goats and sheep. This indication was confirmed in the early years of the Project by the great interest which farmers took in cross-breeding their own local female goats with our project Anglo-Nubian goats (Ch. 4.1.4). Therefore the Project sale prices probably underestimate the value that local farmers placed on the crosses.

Table 4. Average sale prices of goats and sheep from the project farm

	Goats			Sheep		
Price (RO/kg liveweight)	Anglo-Nubian	Local	Cross-bred	Chios	Local	Cross-bred
Price for breeding	1.176	1.200	1.633	1.385	1.136	1.550
Price for slaughter	0.989	1.751	1.519	1.058	1.431	1.179
Overall average price	1.040	1.504	1.557	1.164	1.402	1.434

Source: Taylor, 1991e

4.1.4 Cross-breeding programme with local farmers

When the farmers saw that the cross-bred Anglo-Nubian on local Omani goats were growing better than the local goats, they wanted to participate in the cross-breeding scheme. Initially we were somewhat reluctant to encourage this because we feared that the increased genetic potential of the cross-breds might in fact yield worse results than the local goats if the animals were not being managed with sufficient interest and skill. However, the farmers were insistent and we finally decided to work with them, welcoming the opportunity to respond to their enthusiasm but treating it as a trial and demonstration exercise and insisting that we should be able to monitor the progress of the kids. The contact also gave the staff an opportunity to initiate animal production extension work amongst the farmers participating in the scheme.

After discussion with the farmers we agreed to operate a system whereby they brought their female local goats to the Project's farm. They were kept in quarantine for four days while they were de-loused, de-wormed, clipped of excess hair, hoof-trimmed, given a course of antibiotics and vaccinated against clostridial infections. Any goats judged not to be in good condition were sent or taken back to their owners. After the quarantine period they were penned with one of the stud Anglo-Nubian male goats for a month. Eight to ten females were penned with each male.

While the females were housed on the project farm they were fed a ration of 2kg of freshly cut alfalfa and 0.5kg of concentrates and dates each day, plus an *ad lib.* supply of freshcut Rhodes grass (Barnwell, 1984a). The condition of the animals always markedly improved during their stay on the farm, and this, from the farmer's viewpoint, was an added attraction of the scheme. We regarded it as essential in order to give the foetus a good start.

At first we made the farmers responsible for transporting their animals to and from the farm. However, when the popularity of the scheme meant that we had to introduce a forward booking system, we discovered that the farmers could not be relied upon to bring or remove their goats on the allotted dates. Consequently the Project's extension officer transported the kids, and this gave him the opportunity to see first-hand the conditions under which the goat herd was being maintained, and to give extension advice and support.

The scheme was very popular with the owners of local goats. From December 1977 to February 1983 147 farmers from 19 villages provided 327 goats for a total of 474 services. Initially the number of services was deliberately limited while the impact was monitored, and so most of the work occurred in the period from 1980. Even after 1980 the programme was limited by space available on the Project's farm and by shortage of Project staff time. There was often a 3-4 month waiting list for goats to come in for service.

The programme was terminated in February 1983 because:

- the success of the scheme was proven;
- pen space on the Project's farm was insufficient, as our own sheep and goat numbers had increased;
- staff time could not longer be afforded - particularly for transportation of the goats;
- by then we had a very good understanding of the range of management conditions on owners' farms;
- the Project's own herd was large enough by then to be able to supply selected farmers with good quality cross-bred stock, both male and female.

The Project's monitoring aim was to check the productivity and kid growth rates of the first 100 female goats involved in the scheme, over a three-year period. In practice not all the kiddings in this group were recorded or monitored, because some of the goats were sold, slaughtered, or moved to new areas, or the kiddings were not reported to us. However, 91 of the kiddings were recorded, and they yielded a total of 113 kids of which 44 were twins (24% of kiddings). Some 35 kids were weighed on the day of their birth, and 45 at approximately three months.

It can be seen from Table 5 that the farmers' results compared well with those achieved on the Project's farm. The Project's cross-bred kids of this type were

3kg heavier (26%) than its local kids at three months. So, the farmers also were seeing at least a 26% increase in growth rate over their own local kids. It was this impressive improvement, which the farmers were able to sustain on their own farms, that accounts for the enthusiasm with which the farmers viewed the cross-breeding programme. It has to be said, however, that the participating farmers were almost certainly amongst the better stock-owners. They were largely self-selected - those who showed sufficient interest to find out about the scheme and bring their female goats for servicing. The owners in practice either lived in village houses or on their own small farms. The *shawawi* who owned no land and whose only experience of stock rearing was to rough graze their animals on natural forage were discouraged from participating for fear that their cross-bred animals would fare worse.

Table 5. Weight comparisons of cross-bred goats of local farmers with those on the project farm[1]

Local farmer goats			Project farm goats[2]	
type	number	mean wt	type	mean wt
Birth weights				
single male	12	3.10	single male	3.33
single female	11	2.89	single female	3.03
twin male	4	2.95	twin male	2.67
twin female	8	2.73	twin female	2.37
Mean		2.92		2.85
3-month weights				
single male	16	16.5	single male	17.43
single female	14	13.8	single female	15.73
twin male	6	16.0	twin male	15.37
twin female	8	12.2	twin female	13.67
Mean		14.63		15.55

Source: data from Barnwell, 1984a and Taylor, 1990a.

(1) kids from Anglo-Nubian sires and local dams
(2) means for period 1979-89

Another proof of long-term interest in the scheme was that 36% of the females were served more than once - resulting in up to a maximum of eight kiddings per goat. Furthermore, 15 cross-bred progeny were themselves brought in for mating, for a total of 29 times.

The farmers were similarly interested in the Chios sheep cross-breds with the local sheep. But the Project never accepted farmers' ewes onto the Project's farms (for the same reasons that it had stopped accepting farmers' does). We sold cross-bred stock to the farmers. However, although we monitored their progress, no detailed records were kept. One query concerned ticks. On the Project's farm the cross-breds were reared in a relatively tick-free environment. When they transferred to the farmers' stock pens they were often exposed to more ticks and thus to the disease teilorosis from which they had gained no immunity. Some therefore became sick or died.

4.2 Training paravets

From the date when goats and sheep first arrived at the farm, the succession of Project staff responsible for the health and husbandry of the goats and sheep gave them the necessary vaccinations and remedial treatments as required. Only occasionally was recourse made to the veterinarian based at Sohar. The people in al-Khabura who owned goats and sheep noted the veterinary work undertaken on the Project's farm and started to bring their own sheep and goats (and occasionally cows, donkeys and camels) to the Project for treatment (Cooke and Massey, 1979a-b, 1981a-c). They realised for the first time that such curative and preventative interventions were possible, and more effective than their traditional methods (Cooke and Massey, 1981e). Initially these treatments were undertaken readily as part of the job of the livestock manager. They were relatively few in number, they provided a lot of first hand evidence of the range of diseases, parasites and other problems faced by the local livestock, there was an obvious need to help, and it made a very good point of contact with the stock owners. In some instances, particularly in the case of the bigger animals (cattle and camels) which could not be brought by vehicle to the Project's farm the livestock manager went with the stock owner to his home or farm or, in the case of the *shawawi*, into the coastal plain and the mountain foothills, where the animals were housed. This provided the Project with additional information about the types and quality of stock housing and general management conditions. The equipment and drugs needed for the above work was either bought by the Project or provided by MAF.

As the Project continued to grow, and the goats and sheep on the Project's farm exceeded 200 head of adult stock, it became more difficult to continue *ad hoc* treatments for animals brought to the farm (Cooke and Massey, 1980a-b). Their numbers were growing and their owners becoming more demanding. Often the owners would arrive with their sick animals in the evenings. Some animals were neglected during the working week because the men in the household were working away from home, but when they returned they used their vehicles to take the sick animals to the clinic on rest days. Treating these sick animals was also now of less value to the Project's staff. The initial gains, from the *ad hoc* scheme, of getting to know local disease problems, systems of livestock management and the owners of the livestock, had largely been accomplished. We were by 1980 very well known locally and had many points of contact with farmers and other stock owners.

One approach to helping the farmers (Cooke and Massey, 1981c) was to encourage local farmers to undertake an increasing number of simple husbandry tasks themselves, and train them in the use of basic drugs and simple equipment, which the Project also provided. However, a clinic was also recognised as being essential and the decision was made to hand over responsibility for running it to someone from the village (Cooke and Massey, 1981d; Barnwell and Massey, 1983). By the summer of 1981 a suitable person had been found, Khalifa Ubaid al-Hosni. He had limited formal education, equally limited active contact with livestock and no experience of any form of veterinary medicine. But he had a lively mind, he was obviously practical and he had a quietly confident approach when dealing with animals. At first he worked closely with the farm manager, observing, assisting and asking questions. By the autumn of 1981 he had progressed so well that the clinic was formally handed over to him, and the clinic hours were restricted to four hours in the afternoon, six days per week. In cases of severe illness or of veterinary conditions that he had not seen previously Khalifa sought assistance, but very soon he was treating by himself almost all the animals that were brought to the Project's farm. Anyone who arrived outside clinic hours was asked to return the next day or was directed to Khalifa's house where he ran a small private clinic as well. The most complex and/or unpleasant cases that Khalifa dealt with were difficult births (including kids or lambs that had died in the uterus), appalling cases of flystrike (where maggots had largely separated the skin from the flesh) and such things as broken bones and deep wounds inflicted by dogs or barbed wire. He took all these cases in his stride, and almost always with success. In addition he vaccinated the Project's sheep and goats and dealt with a wide range of viral and bacteriological infections.

The clinic in which Khalifa worked initially had no facilities other than a store for basic equipment and a refrigerator for drugs. Animals were examined and treated in the yard that also acted as a parking place for the cars and pick-up trucks that brought the animals to the clinic. In later years the clinic's facilities always remained simple but they were gradually improved. The main alterations included laying a very smooth easy-to-clean concrete floor, and providing water and a drain. The floor space was also covered with a simple open-walled shed and a roof to provide shade. Thus Khalifa could work out of the sun, and keep the clinic clean and disinfected. The animals also benefited by being treated in the shade, while the owners helped the work by keeping them still.

From the day that Khalifa first took responsibility for running the clinic, he also undertook to record in a register every animal treated. He took pride in this work, which meant that the register was carefully maintained for the full six years, 1981/2-1986/7, that he was in charge. Essentially each record was a one-line item, including species, sex and age (young or adult) of animal, date, diagnosis, treatment and charge. In the later years the record also included the name of the settlement and the *wilaya* from which the animal came, and the name of the owner. The register had limitations including gaps when Khalifa was on leave. Also, Khalifa's limited diagnostic skills and understanding of English meant that diagnoses were often generic in nature. But though this record would not have been acceptable from a qualified veterinarian, it tells us what can be achieved by a 'barefoot vet' with no formal training, given a supportive working environment.

When the clinic was first mooted it would have been the only one in al-Khabura. However, by the time it opened a veterinary assistant had been appointed to the staff of the local agricultural extension centre. He worked the standard government hours from early morning to midday, which is one reason why we restricted the Project's clinic to the afternoons so that the two clinics could be seen to complement each other. The veterinary assistant was competent but he was not always well supplied with drugs and veterinary equipment by MAF. He was an expatriate, from India, but as our work made increasingly clear it was perfectly possible to find and train young Omanis to do the work and cope very satisfactorily with 80-90% of the cases seen.

The Project's experience with Khalifa poses important questions. His training was quick and cheap but did it produce someone of real worth? Could other people valuably be trained in the same way? Was keeping the register a worthwhile exercise?

The register informs us that Khalifa dealt with 31,000 animals in six full years, and that the number of cases rose steadily each year (Dutton, 1994). This confirms that the livestock owners had a growing respect for Khalifa's work. One concern, at the outset, was that many of the older people might not accept treatment for their animals from someone who was not only unqualified but also very young. However, neither Khalifa's youth nor his lack of formal training troubled anyone. Khalifa was judged entirely on his results. The growing number of cases treated probably also reflects the growing awareness by the local population that things could be done to cure livestock illnesses. The sense of fatalism that had inevitably characterised the local response to livestock illnesses in the days when no veterinary assistance was available, was slowly breaking down. Indeed some members of the community were becoming quite aggressive, demanding that their sick animals should be treated. This reflected the fact that livestock were (and remain) very valuable assets. Whilst the indirect effects of the oil economy have greatly reduced the value of some agricultural produce, such as wheat and some varieties of dates, the value of local livestock has grown (as shown in Ch. 4.1.3), even in spite of the competition from cheap imported sheep from Australia.

The register also shows that the proportion of cases occurring in November and December was consistently low and that the monthly proportion thereafter increased gradually through to February and then rose rapidly to peak in March and April. The winter low is presumably associated with low temperatures reducing the likelihood of any heat stress and thereby reducing disease susceptibility. The peak in March/April will be related to such factors as moist conditions associated with late winter to early spring rains coupled with increasing temperatures and a rapid increase in the number of flies and therefore fly-borne diseases and other veterinary conditions (including fly-strike). The dry heat of May and June kills the insects which probably accounts for the drop in the number of cases coming to the clinic through the summer months (see also Kwantes, 1994).

Almost 86% of the animals treated were goats. Sheep formed only 13% of the total. Goats in fact form a high percentage of the livestock on the Batina, however, as noted above, a detailed survey of domestic livestock in the period 1973/4 revealed that there were four sheep to every six goats (Dutton, 1982f). If this goat/sheep ratio remained unchanged in the 1980s (which is likely) it suggests that goats are more susceptible to disease and other veterinary conditions than sheep. This was in fact also strongly indicated by the veterinary record of the goats and sheep on the Project's own farm (Taylor, 1991a) and at the Wadi Quriyat Animal Breeding and Applied Research Centre (ABARC) (Kwantes, 1994). The susceptibility of goats was also noticed in the years 1980/1, before the Project's clinic opened. Cooke and Massey (1981c) noted that the: 'difference between goats and sheep has been consistent throughout the period for which the Project has been active at al-Khabura and reflects a real and important difference' in disease tolerance between sheep and goats' (p. 8).

Khalifa recognised some 60 different conditions. Some were partly synonymous (e.g., worms and stomach worms, or boils, swellings and ulcers) whilst others are generic terms covering more than one disease - notably pneumonia where, for example, no distinction is made between pasteurella and contagious caprine pleuro-pneumonia or other upper respiratory tract disorders. The most common conditions, in descending order, were pneumonia, digestive tract disorders, abscesses and other swellings, breeding and birth problems, external parasites and flystrike, mastitis, anaemia and dermatitis. Sheep were significantly more susceptible to abscesses and external parasites, goats were more susceptible to mastitis, breeding problems, foot and mouth disease and PPR. Pneumonia (loosely diagnosed by the presence of a rumbling chesty cough and a nasal discharge) was the largest single problem with which Khalifa had to deal, accounting for 32% of total cases. Incidences rose to a sharp peak in March and April in the hot but still moist conditions and when flies were abundant, and fell to a low in the very hot, dry month of June. Comparison with rainfall graphs suggests that the peaks of pneumonia followed rainfall. However, there were always cases of pneumonia present. It was an endemic and debilitating condition which flared into prominence when the environmental conditions favoured it.

Thus Khalifa's register, simple though it was, tells us firstly that a veterinary assistant without formal training was able to maintain a complete and informative record of his activities, even in a foreign language. The main points are worth reiterating:

- identification of the range of problems found;
- clear indication of those which were most common and severe;
- differentiation in disease susceptibility between sheep and goats;
- indication of the seasonal distribution of diseases;

- correlation between some diseases and climate factors (notably pneumonias in the spring months).

All of this information is of value in planning the future expansion and development of the veterinary services as a whole.

Also, during the six years that Khalifa ran the clinic he fully convinced successive senior Project staff of his competence. He was reliable, handled both the livestock and their owners with sensitivity, looked after the drugs and veterinary equipment with care, and thoroughly cleaned the clinic after each session. Khalifa had innately strong qualities for the work. He was naturally practical, he had a good empathy for animals and their welfare, he was methodical, he was not in the least squeamish when dealing with blood and gore, and he was very willing to learn from advice and by example from others. But to judge from his poor school record, Khalifa was not academically bright, failing his primary school examinations more than once. He is therefore a convincing proof that formal academic success is not necessarily essential in order to make a major contribution as a veterinary assistant. But his working and learning environment was very important to his success, and the key elements of this were as follows:

- a large throughput of animals for treatment - on average over 20 per day;
- a regular supply of appropriate drugs and equipment;
- a simple but fully adequate clinic in which to work: easy to clean floor, piped water and shade;
- senior staff to give advice and help when necessary;
- endless encouragement and enthusiasm openly expressed by the senior staff.

With this support and encouragement Khalifa greatly exceeded our original expectations by the success with which he also trained his Omani assistant. The competence of the assistant added to our conviction that it is possible to find potential paravets within village communities and provide on-the-job training which converts them rapidly into very valuable practitioners.

The clinic also demonstrated that people were prepared to pay for the services they received. The services were charged (though not the economic cost) and there was never any objection to paying. Payments were levied not only to increase the budget for more drugs and equipment (and so maintain the clinic sustainable) but also to encourage livestock owners to take more care of their animals in order to prevent them becoming ill. The main problem about running an almost entirely subsidised service (as MAF still does) is that it becomes financially more sensible - from the narrow viewpoint of the livestock owner - to give minimal care and management to the animals and then rely on the government to pay the cost of curing them when they fall ill. This exposes the government to budgeting ever higher allocations to maintain the veterinary service in operation (money which, in the longer-term, the government will not have) whilst reducing to a minimum the sense of responsibility that a livestock owner has for the well-being of his stock.

Sadly, Khalifa's lack of any educational qualifications meant this his expertise could not be officially recognised. Although it was widely accepted by officials within MAF that he was excellent at his job, and that his training was extremely cost-effective, no serious attempt was made by the bureaucracy to find a means of giving him formal employment, or even recognising his own small clinic. An applied, practical series of tests would have been sufficient to reveal his capabilities in comparison with others, and indeed to reveal perhaps where he would have benefited from additional practical training. The fact that Khalifa, and the assistant he trained, were so able, merits giving this issue full consideration. There must be many potential Khalifa's in the villages of Oman.

In 1987/8, as part of the process of integration between the Khabura Project and MAF, it was agreed to move the clinic to the al-Khabura extension centre and for the Project to have some degree of overall responsibility for the government clinic as well as our own. The former would continue in the mornings only, the latter in the afternoons. Andrew Gauldie, the Project's veterinarian, organised changes which improved the physical layout of the clinic. A practical problem was a reliable supply of equipment and drugs. The Khabura Project's small budget for these items was supplemented, when the clinic was run on the Project's farm, by increasing the charges to the farmers and buying new drugs and equipment with the money thus raised. We had even opened a special veterinary account at a local bank, and this was operated by Khalifa and his assistant, Abdulla, under Andrew Gauldie's general supervision. However, making such charges was contrary to MAF policy. The issue was raised with MAF veterinary officials. They were sympathetic to our predicament but unable to state clearly whether or not we would be able to continue charging in order to maintain a good service.

4.3 Dairying

As anticipated, it soon became apparent that one consequence of maintaining the Project's sheep and goats at a consistently improved level of nutrition was a higher yield of milk. General observation also indicated that the cross-bred female progeny, from locals crossed with the Anglo-Nubian goats, produced even higher yields of milk. At first the Project was content to allow the kids to suckle all this extra milk in order to promote healthy early growth before weaning. However, we also had the scientific curiosity at least to quantify the milk yields of different local and cross-bred goats, and sheep. We also wanted to measure the yields of some of the local cows, which were kept principally as milk animals, so that comparisons could be made (Horton, 1984a). As far as we were aware, no one previously had measured the milk yields of any of the domestic ruminants in northern Oman so our data would be of primary value when consideration was given to small-farmer dairy development projects based on cows or on goats and sheep. We also sought information, based on our own first-hand experience of handling the animals, about suitable designs for a milking parlour and dairy for goats and sheep, and about the costs of milk production and the likely income from sales of milk and milk products.

4.3.1 A Project milking parlour and dairy

To meet the above objectives for goats and sheep we envisaged testing a small-scale dairy, built on the Project's farm, with a throughput of milk from a sub-set of the Project's sheep and goat herd. From the outset it was envisaged that if a goat-sheep dairy enterprise looked worthwhile we would establish a trial dairy herd on a local farm (if we found someone who was seriously interested) and process his milk through the Project's dairy. The dairy might then become a village dairy centre, either to process and market milk and milk products itself or to act as a collection centre for a larger sub-regional dairy. A general argument in favour of goat's milk is that it is not only highly nutritious but also more easily digested than cow's milk which makes it of especial value to families which include young children and/or elderly people (Devendra and Burns, 1983).

Obtaining milk yields and management information from cows entailed, initially, becoming friendly with a number of local cow owners and visiting them regularly when milking and other cow management operations were taking place. This was a time-consuming procedure but had the inestimable benefits of yielding accurate data about yields per day and per lactation, and about feeding, breeding, rearing, housing and other management practices, and about illness, mortalities and changing veterinary inputs.

A potential attraction of the dairy project for the small farmer was a fairly dependable (even if not very large) daily income. In contrast, for the few people who then raised animals to sell for meat, the sales realised only an irregular income, once or twice a year, from animals which were quite likely to die from disease before they got to market. Most people, however, raised goats and sheep solely for home meat consumption to celebrate one of the religious festivals. They preferred a daily expenditure of money on foodstuffs in order to avoid the single large purchase price of an animal at the *Id* sales. For both groups of owners it would be much easier to budget if they had a regular daily income from milk, and to be able to set this income against costs.

But a potential danger of the dairy project for the small farmer lay in its potential attraction. If he marketed his milk without due regard for the proper nutrition of the kids and lambs, the latter would become stunted and gradually undermine the overall quality of his herd.

On behalf of the Project, the first person to tackle these issues was Mrs Lynne Horton (1984b, 1987) who had excellent previous experience of running her own small goat dairy business in Britain. She was also a woman, and although this was immaterial as far as setting up the Project's dairy unit was concerned, it was a positive asset when collecting information about the cows which were mostly managed by the women in the local village and farm households. Generally they were very pleased to have the opportunity to chat with a woman from another country, and they readily provided all the information asked of them and even let Lynne handle the cows.

There were three local situations for which goat dairying (or sheep dairying if the sheep proved to be a suitable dairy animal) might have played a valuable role in improving the nutrition of the farmer's family, or in increasing family income. Two of the situations were non-commercial, the first for a family without a house cow. In this case three higher-yielding goats would be segregated, in order to ensure a higher level of nutrition and prevent random breeding, and mated at four-month intervals in order to achieve a continuous and reasonably even milk production throughout the year. Assuming that one goat is dry and in kid, a second at full production yielding 3 litres/day, and the third at half production yielding 1.5 litres/day, the family would obtain 4.5 litres/day throughout the year. This would compare very favourably with the output from a house cow, which at peak production would yield a maximum of 7 litres/day (often less), falling gradually to 2 litres/day followed by a 2-3 month dry period. The second non-commercial situation envisaged a family with one house cow. One or two goats would be mated in order to kid as the cow neared the end of its lactation and thus be in milk when the cow was dry.

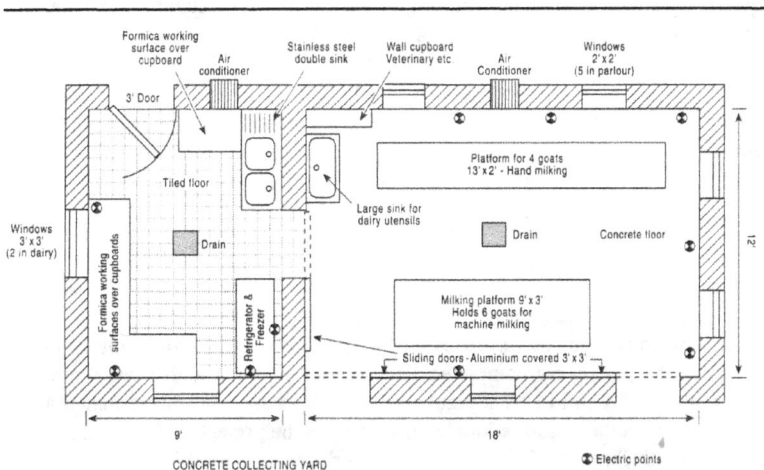

Source: Horton, 1984c

Figure 8. Ground plan of dairy and milking parlour

The third situation envisaged, a commercial enterprise, was the one under examination on the Khabura Project farm. Here the enterprise was first established with a herd of 12 does. Milk yields and food intake were closely monitored to provide information for a cost-benefit study, and an intensive effort was made to develop and refine a production and management system that would optimise return on financial and labour inputs.

The Project also designed and built a simple, but air-conditioned, two-room building for the parlour and dairy (Figure 8). The parlour was equipped with locally made milking stalls to accommodate ten goats or sheep, weighing scales and a blackboard for milk recording, a veterinary cupboard, a sink and a drain. Two doors to the pens allowed the goats to circulate in one and out the other. Another door led into the dairy where were a double sink, a refrigerator and freezer, cupboards and working surfaces. It also housed simple milk processing equipment for making yoghurt and for packaging yoghurt and milk. Nothing was complex, everything was basic and easy to clean, but large enough to handle milk from 10-20 local producers if the dairy programme ever reached that stage. The dairy unit was in operation by the spring of 1984, and formally opened by the Wali of al-Khabura on 15th May in the presence of HE the Minister of Agriculture and Fisheries. At a small party to celebrate the event, liberal quantities of goats' milk yoghurt and cheese cake were enjoyed by all.

When the dairy was in operation and the Project's dairy herd in full production, a trainee, Khalifa bin Khamis, was taken onto the staff to work alongside and learn from Lynne Horton. He proved to be conscientious and quick to learn, and his quiet handling of the stock was outstanding. He became proficient in all aspects of the work from feeding and watering the dairy animals and cleaning pens, to all the activities of the milking parlour and dairy. He also performed basic veterinary routines such as foot trimming, worming, and treating eye infections, cuts and other minor ailments. In fact his level of interest and skill developed so rapidly that he expressed strong enthusiasm for having the first off-project trial and demonstration unit installed on his own farm.

4.3.2 Small-farm dairy enterprises

Khalifa's dairy goat pens, solidly built in December-January 1984/5 to a design drawn up by the Project in discussion with him, incorporated a milking platform with four stalls. The pens, partially roofed and partially cement floored, with overall measurements of 40 feet by 20 feet, cost only RO475 including labour.

When the pens were completed Khalifa was supplied with three cross-bred goats, all heavily in kid. They gave birth in February 1985, were mated at the end of July with a cross-bred male on loan from the project, and kidded again at the end of December. The six kiddings produced 14 strong kids with a weight range of 2.7kg to 3.9kg, thus giving an excellent start to Khalifa's on-farm goat dairy project trial. Between February and October 1985 the three goats yielded, respectively, 430kg, 475kg and 480kg of milk. Some of the milk, supplemented with milk replacer, was fed to the kids, some was kept for domestic consumption and 786 litres were sold to the Project's dairy at 300 baisas/litre which gave Khalifa an income of RO236. Khalifa's routine in both lactations was to sell the morning milk to the Project and feed the afternoon milk to the kids or, after weaning, keep it for the household.

Unfortunately, from the point of view of monitoring and information gathering, the first batch of kids had to be reared on Project's farm because Khalifa's pens were unfinished. However, the six kids in the December batch were all naturally reared on their dams, with additional fodder and concentrates from an early age, and weaned at 12 weeks. The two best females were kept as followers and the remaining four kids were sold for RO45 each.

In summary, Khalifa fed his animals well and looked after them with care, and these good management practices were reflected in high milk yields and rapid kid growth rates. He was also able to sell his kids at a young age for a high price, to receive a small income from milk sales and to provide his household with milk after the kids were weaned. The value of the milk yields was well in excess of the feeding costs.

Steps were also undertaken to initiate other small enterprises. In February 1986 one goat (with triplets) was sold as a supplier of milk to a family with no house cow. The kids in part were bottle fed and in part were suckled from their dam. They grew well and, after weaning at 12 weeks, the milk was used by the household. Also, in February 1986 a dairy goat and her twins were sold to another household with a house cow. These kids also grew well, and on weaning the milk was used to supplement that from the house cow.

Zawan Rashid, a local woman, replaced Khalifa Khamis as dairy assistant. Zawan was unusually suited to the job by having few family responsibilities (separated from her husband and with two grown-up children), familiarity with both English and Arabic numerals, and a strong, almost overpowering, confident personality She was a unique woman in this rural community in that she owned an old but powerful car which she regularly drove to and from Dubai. The car allowed her to take responsibility for delivery and marketing of dairy produce, and her strong personality allowed her to enforce rigorous standards of hygiene which we, as outsiders, would have been hesitant to insist upon. Zawan was industrious, learned the dairy routines very quickly, proved to be an invaluable assistant when visits were undertaken to obtain information from the cow owners, was confident about taking responsibility for the dairy when Lynne Horton was on leave, and, as proof of her commitment, purchased two cross-bred dairy goats from the farm as 'house goats' to provide her household with daily milk. She lived in the middle of al-Khabura village where there was no room for a cow, but she kept the goats successfully and cleanly in a shelter built on the roof of her garage! And although one would not normally recommend a garage roof as a goat pen, Zawan demonstrated the useful role to be played by house goats in restricted circumstances.

By 1985 the Project's basic dairy goat management system had been established and was employed throughout that year, yielding very satisfactory results. The feed was a combination of green forage (alfalfa and Rhodes grass) and concentrates purchased from the Oman Flour Mills factory together with some dates when locally available. Up to a maximum of 3kg forage were fed daily to each goat, with a much higher proportion of Rhodes grass in the summer months, when it was more readily available. Per head the daily concentrate range was 0.15kg-1.5kg, the exact amount dependent on stage of lactation and dry period, milk production, size (the small Omani goats received less), and parity (young first kidders which were still growing were fed more). The routine stock work included monthly weighing and foot-trimming, an after-kidding routine (including worming, hair clipping and washing) and an early summer clipping. The general veterinary treatments were for worms, scours, mastitis, respiratory disease and loss of appetite. Checking the animals' health and condition was part of the daily routine undertaken at feeding time or when milking.

The Project's goat herd and dairy were both in full production and local markets for milk and yoghurt had been established. In this manner the dairy project continued until the staff transferred to other centres in mid-1989. But the most comprehensive cost-benefit analysis was made for the full year of 1985 (Horton, 1987). The Project had enough data to prepare a detailed economic study of the dairy goat herd on the basis of actual costs and prices. The costing was made on the basic assumption that the herd was owned and managed by a local Omani family employing only family labour, therefore the costs did not include project labour nor the dairy specialist's time. The feed costs included the actual concentrates consumed in 1985 and an assumed 3 kg/day/doe of forage at a production cost of 25 baisas/kg. The milk income was based on a figure of 374 baisas/litre which was the selling price of pasteurised milk after deducting packaging costs (yoghurt made a higher profit of 400 baisas/litre).

On the above basis the total income during 1985 was RO2,197 and total costs were RO735, giving a profit margin of RO1,460 from 12 does. This is a conservative figure because 3kg per day was the maximum not the mean forage consumed, and the likely production cost of forage was only half or less than the 25 baisas/kg assumed[12].

In addition to the feed, the veterinary costs for the whole herd totalled RO16, while cleansing fluids and udder wash totalled less than RO35, some RO50 all told. On the other hand, on the benefit side of the analysis, 23 kids were born to the dairy herd in 1985. Their added value at one year of age, representing total value less costs, was over RO30/head, a total of RO690.

Based on the above figures a herd of 20 does would yield a net income of RO3,130, or RO2,430 if RO700 of labour costs are included in the calculation. This represented an attractive annual income in the mid-1980s, and about three times the value of the Rhodes grass if the 12,840kg had been sold at 250b/4kg bundle.

---

[12] Taylor (1991e), on the basis of three separate estimates, assumed a cost of RO5/tonne (wet) for home-grown Rhodes grass, equivalent to 5 baisas/kg.

When the dairy project started there were few if any formal regulations concerning dairy hygiene or marketing procedures. We adopted our own common-sense rules within the constraints of our financial circumstances and availability of packaging materials for dairy produce. In practice, the produce never caused any health problems. However, at the end of 1985 new GCC validated laws were introduced concerning stages of processing, food handling, packaging and date stamping. The Project could comply with most of the new regulations, but not with the laws demanding that production by fully mechanised, nor that processing should be limited to 'licensed plants' which had to be business registered. There was nothing wrong in principle with the new rules but they started with the assumption that all dairy work would be undertaken by large, mechanised and fully commercial dairies. No one had envisaged the development of a village-scale industry built from the basis of traditional, local understanding of dairy hygiene requirements. Nor had they considered a dairy which would act as a centre through which new ideas and new standards would be actively disseminated to the farmers because its success would be inter-dependent with theirs. The new rules made the building 'up' of a dairy industry from a truly rural base, very difficult if not impossible. During 1986 and 1987 the then dairy staff (mainly Alison Sherwood and Kate Rogerson) tried to argue against these regulations. In general we had the tacit - and occasionally active - support of MAF but responsibility for implementing the GCC regulations lay with the Ministry of Commerce and Industry (MCI). In practice we were not deemed to be sufficiently important for them to bother with. No one in MCI seemed interested in small rural enterprises.

It gradually emerged, also, that the new rules would not allow the dairy to be in the same building as the milking parlour and therefore in close proximity to the livestock. For the Project's interest in dairy work to continue it would either have to build another dairy on the property further removed from the goats and sheep, or establish a new dairy in a rented building in al-Khabura new-town. We decided to explore the latter option, partly because there was not sufficient room for a new dairy building on the Project's farm. More positively we also realised that if a village dairy was opened in the new-town it would be much more part of the community, and it would be more widely known about and more accessible than on our property to both suppliers and customers. In practice we had the choice of several empty commercial properties which, with fairly non-costly modifications, would satisfy the new dairy regulations.

When the decision was taken during the winter of 1988/9 to close the Khabura Project in mid-1989 and transfer its specialist staff to the national livestock research stations, it was no longer sensible further to explore the option of opening a small dairy in al-Khabura new-town. Data on the Project's dairy herd continued to be collected until the station closed, and the dairy herd was moved to Rumais.

Unfortunately it has to be reported that for reasons which neither we nor his relatives understood Khalifa's personality appeared to disintegrate after 1986, to the point where he became dissolute, with the sad consequence that he effectively lost interest in his dairy herd.

4.4 Spinning and weaving

The spinning and weaving project, which started in al-Khabura but later extended to Rustaq, employed four expatriate advisers 1977-89. Although some goat hair was spun, the main source of raw material was wool from the local sheep. A mixed collection of fibre samples from al-Khabura sheep, and others from Oman and nearby Fujairah and Buraymi, were described as 'carpet-wooled rather than "hair" sheep, suitable for rugs, which you are making'[13].

4.4.1 Initial years with MAF

The proposal to MAF (Dutton, 1977) to initiate a spinning and weaving project pointed out that during the 1970s it had become less financially worthwhile for the spinners and the *mansul* weaver in al-Khabura to continue their work - it was very labour intensive and the financial return small. In consequence 'this potentially remunerative craft industry is in danger of dying, and the wool (an important animal by-product) would then be entirely wasted.' The project benefits were then seen as threefold: the effective use of a major animal by-product, wool; the provision of rural employment for both men and women; and an increase of total rural income based on a locally available agricultural raw material. Also, it was the only way at the time that the author was able to incorporate women in the Khabura Project in a productive enterprise so that they as well as the men might play a role in rebuilding the productive mutual self-reliance that the Project was seeking. The initial project was specifically called a 'pilot project' to run for one year to see whether local people retained any real interest in working with wool in the new oil age. If successful its findings would also allow the next stage of the project to be more closely defined.

Fieldwork started in the inland (*shawawi*) part of the village of al-Hugayra, which is where the village livestock research survey had shown that a large number of women retained an interest in their spinning skills even though the male weaver in Rudayda, for whom they had spun and plied yarn, had stopped working in 1976 (Dutton, 1982f). The women in 'inland' al-Hugayra and, subsequently, the other 'inland' villages on the Batina, were regularly visited by successive spinning and weaving advisers. Women also started, as early as 1977/8, visiting the adviser in the Khabura Project's weaving workshop to bring yarn and woven goods and to discuss the work they were doing (Crocker-Jones, 1979).

---

[13] Dr M.L. Ryder, ARC Animal Breeding Research Organisation, Roslin, Midlothian in a letter to the author dated February 1980, in which he also said that if we wanted to go for finer wool we should select what he called the hairy Shetland type of local sheep.

Early attempts were made to introduce the exotic technologies of the spinning wheel and the floor loom. The women spun many kilograms of yarn on the imported spinning wheels (Dutton, 1979) but in practice the spindle with which they were very familiar (Figure 9) retained several advantages over the wheel. First, it was costless to make and renew, occupied no space and could be kept out of reach of the children. Second, it was equally easy to use whether sitting on the floor or a chair, or standing or walking or riding in a vehicle. It could therefore be used anywhere around the home, when visiting friends or following the sheep and goats. In practice, also, the spinning wheels were not sufficiently robust to withstand the rigours of the climate or rough treatment from over-inquisitive children and goats. Wheels could, however, in the longer-term future, have a bigger role. People skilled in the use of wheels can produce good-quality yarn faster, which may become an important factor. Also, by the end of the project's life more women were spending more of their time in environments more friendly to spinning wheels; block built houses, with firm cement floors and even air-conditioning, from which domestic livestock were mostly excluded.

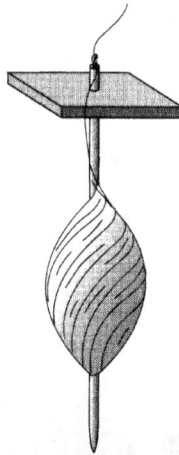

Source: Lochhead, 1983

Figure 9. Traditional spindle typical of northern Oman

The imported floor loom was very expensive, and it was not thought likely that the women would want to use it, which, in general, proved to be the case. In its working principles, however, it was not dissimilar to the pit loom formerly used by the male Rudayda weaver, and would have found a use for making a wider range of broad single-strip *mansuls* or rugs if a male weaving apprentice had been found. The floor loom occupied much less space than the pit loom (because the warp was wrapped around a beam) and could stand on a firm floor. It could have been attractive partly because the weaver would have worked in a cooler indoor environment. However, no male (or female) weaving apprentice was forthcoming, other than a reluctant man whose interest waned in the hiatus between the first and second advisers. But the loom was very suited to the weaving workshop, it was a technology with which the Project's weaving advisers were familiar and in particular it allowed the first of them, Gigi Crocker-Jones, to become fully familiar with working the yarn and to demonstrate a wider range of products to those women more interested in experimentation and design.

But all the woven products had to have markets. A small local village market still existed, but it was in decline. The male weaver had only woven to make *mansuls* for the women who span the yarn. He had never attempted to find other markets or increase his product range. However, when he stopped weaving the local women had to find alternative sources of rugs or mats. In practice, typically, their husbands bought them in the Gulf, where they were working. They were imported from overseas, more colourful, mass produced and cheap. Therefore the local market for *mansuls* was never likely to be restored. If the weaving project was to become sustainable it had to create markets for new products, and a marketing system.

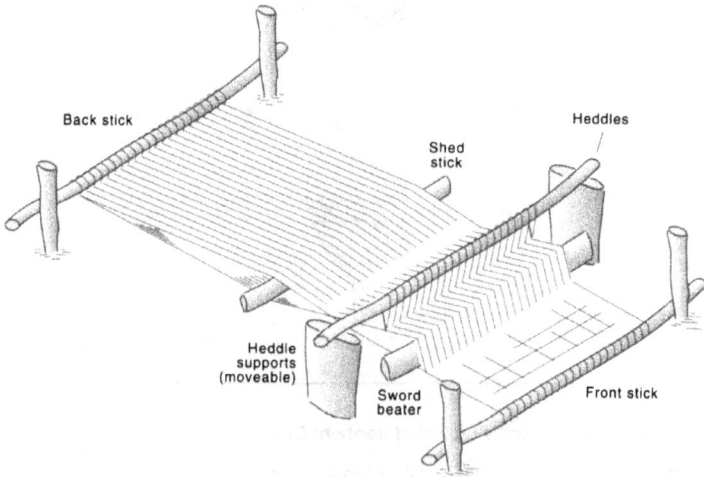

Source: Lochhead, 1983

Figure 10. Traditional ground loom, as used by women on the Batina

With markets in mind, the initial (and on-going) spinning priorities were to obtain a high average quality of yarn and to encourage the production of a greater range of types of yarn suited to a greater diversity of woven products.

92

The local women expanded the range of straps and other goods that they had previously woven as donkey and camel trappings. They also produced on simple ground looms (Figure 10) wider strips of cloth (up to about 40cm) which could be sewn together to make high quality rugs, bags (Figure 11a-b) and wall hangings (Brokensha, 1980). Early progress with marketing was made. An initial successful sale of rugs was held in June 1978 in Muscat, followed by another sale in December 1979 (Brokensha, 1981). By then there was also a small but growing demand by expatriate women in the Muscat area for the yarns being produced in al-Khabura. The women used these yarns for weaving, knitting and macrame work. Some 200kg of yarn had been exported to Britain, and contacts had been made with other possible yarn purchasers in Finland, Germany and Australia.

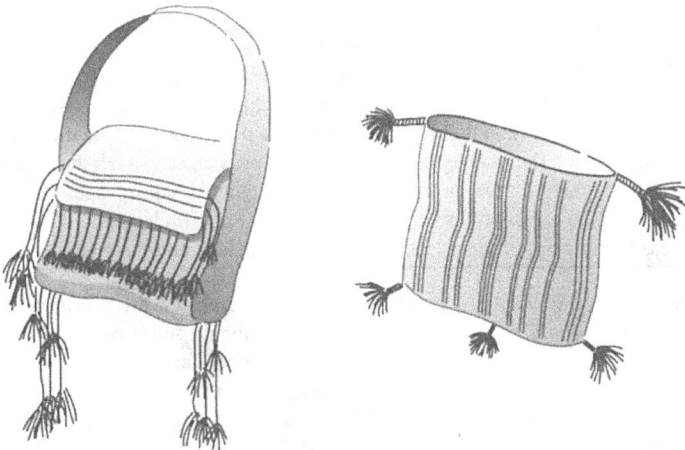

Source: Lochhead, 1984b

Figure 11a-b. New product designs for bags

4.4.2 With the Ministry of Social Affairs and Labour

Unfortunately there was a 10-month gap between the second adviser leaving and the third one arriving, occasioned mainly by the cabinet decision that MAF did not have rural crafts in its remit. After lengthy discussion the Ministry of Social Affairs and Labour (MoSAL) agreed to sponsor and fund the project. Because the new weaving adviser reported to, and remained in on-going negotiation with, a different ministry than the rest of the Khabura Project, an organisational divide was created though the adviser continued to live on the Project's farm and fruitfully interacted with the Project's livestock staff in discussions about sheep and with the engineer in the design of modified looms and other equipment. Beneficially, however, the link in MoSAL was with its community development work (Heath, 1996) which encouraged the advisers to develop their natural interest in women's welfare including organisational skills and literacy.

In writing about the changing response of the women up to 1984, Alison Lochhead (1984a), the third of the four weaving advisers to participate in the project, wrote:

'Initially, the women only produced for the money they earned. They took no pride in their achievements, they had no concept of the need for high-quality work in relation to market need, they were unable to come together in groups for training, or to concentrate for extended periods, or to take any responsibility for organisation. They relied totally on the weaving adviser for supplies and motivation.'

However, she was able to add that by 1984:

'...the women in the project have made enormous progress. They now organise their own supplies, are coming together in groups to learn, are beginning to train each other, and can concentrate for long periods. They now take a real pride in their achievements, are using their imaginations and enjoy discussing the future possibilities for the project, for which they take a much more active role in running. The women's confidence in their abilities has increased greatly. Their work is now of a high standard with an increasing range of products which are highly saleable and in demand.'

The value of pride was also explained by Charlotte Heath, the fourth weaving adviser. Even more important than money the 'innate sense of pride in their work is more valuable and productive than any other factor; it ensures skilled work, high quality and independence.' (Heath, 1985a).

In the period 1982-4, as described below, the spinning and weaving project also made progress in terms of local staff, location of field bases and development concepts. It also made progress in turning to advantage the new link with MoSAL. But there were frustrations, particularly with marketing, and there was also a downside to the MoSAL link.

Project expansion necessitated the appointment of an assistant to take responsibility for the basic tasks of administration and buying and selling, while allowing the adviser to concentrate on training and broader project developmental issues. The adviser decided to find an assistant from the al-Khabura area who was literate. Discussions were held with the women about the assistant's role, and they were asked if they knew of a woman, or husband and wife team, who could fill the post. The women concluded, however, that at least initially the assistant would have to be a man. The only women who could read and write lived in the villages on the sea shore and they looked down on the *shawawi* women of the coastal plain villages and spoke of them, in a derogatory tone, as being 'bedu'. No women from the *shawawi* households had the confidence, and perhaps freedom, to move easily outside the locality or to travel to the capital on project business with MoSAL and handle petty cash. A male assistant was therefore appointed, in September 1983. He ran the workshop on the Khabura Project farm as a place where: farmers could sell their fleece, spinners could buy fleece and sell yarn, and weavers could buy yarn and sell woven products. He was trained in all aspects of handling the spinning and weaving equipment and running project finances. His appointment gave the women greater confidence in the future of the project, and spinning and weaving production both increased. Happily, the women did not become over-dependent on him but continued to organise their own supply networks, using him as a last hope resource or to obtain different types of yarn that came from further afield (Lochhead, 1984a). The assistant was sent on a training course by MoSAL but thereafter although, potentially, he had a key long-term role, his brief and loyalties were somewhat divided. In addition to him, the advisers encouraged volunteer women helpers, starting in 1985 and 1987. Two were on the MoSAL payroll by the time the last adviser departed in 1989 (Heath, 1996).

When, under MoSAL, responsibility for the weaving project moved to its Director-General of Social Affairs in January 1983 it also became part of the National Community Development Programme (NCDP). By this time the weaving adviser was sufficiently confident about progress at al-Khabura to think of opening a new centre. Logically it was appropriate for this to be located somewhere accessible to al-Khabura, in a place where there was a strong tradition of spinning and weaving and where NCDP could provide additional local support within a broad community development context. The region of Rustaq in the foothills of the Hajar mountains was selected. NCDP already ran classes for women in the area (in aspects of agriculture, health and child care for example), and so group learning for weavers seemed feasible. Alison Lochhead met and spoke with women from several villages in different *wadis* near Rustaq before settling on the village of Ayn Amq in Wadi Sahtan. In practice the Rustaq based staff of NCDP were not able to provide much effective support. Nevertheless, a protected training area for the weaving project was finally erected and training started in November 1983. Needing yarn, Alison Lochhead visited all known spinners (who were male) in the local *wadis*. The response was overwhelming. Due to the superiority of the fleece the mountain yarn was much finer quality that the yarn from the Batina. Yarn surplus to the requirements of the centre at Ayn Amq was sold in al-Khabura. The response of the spinners clearly showed that with encouragement and a market demand the traditional spinning craft could be revitalised. But of course for long-term sustainability, the demand for woven products would have to grow as well.

By May 1984 some 70 women near al-Khabura were weaving regularly or intermittently, and 160 were spinning. In the Rustaq area 17 were being trained and approximately 50 spinners were supplying yarn. The products are listed in Table 6, together with the then prevailing prices per kilo or per item. All work was part-time, fitted in between domestic duties. Earnings, therefore, ranged widely, from RO3 to RO100 per month. The women kept all the money themselves. They used it to buy a mixture of luxury items such as perfumes, and also clothes for their children, child education, and medical requirements. The significance of the income can be judged by the fact that the average locally-earned male wage (for the relatively few people who could find such wages locally) in the al-Khabura area was RO120 per month. If the family was earning no outside remittances, the weaving income relieved financial pressures.

Table 6. Spun and woven products, and their prices, May 1984

Item	Price (RO)	Detail	
Spun yarn	3.0/kg	plied	
	2.0/kg	unplied	
Rugs:	9.25/kg	6.00/kg for labour	
		3.00/kg for spun wool	
		0.25/kg for yarn wastage on the loom	
an average rug	15.725	weight - 1.7kg	
		therefore:	10.20 labour
			5.525 spun wool
Bags	3.00	simple design, small size	
	15.00	complex design, large size	
Cushion covers	7.00	simple design, small size	
	15.00	complex design, large size	
Camel straps	2.50	black finger-braided, 8-foot length - embroidered ends	
Decorated girth straps	3.50		
	5.50		

Source: Lochhead, 1984a

At al-Khabura training, initially, was one-on-one. The women were uncomfortable with the idea of coming together in groups. Eventually they agreed, and they found that they liked the social experience. Groups were not only a much more efficient way of using the adviser's time, there were other benefits also. The women helped each other. Women who had refused to train others began to try. Their concentration time increased enormously. They were even putting pressure back on the adviser, expecting her to be prompt when a group meeting had been called for a particular time. A hierarchy of ability, keenness and openness to innovation became manifest. Some women made use of pattern books and started to make tapestries incorporating facets of village life and mountain scenery. In groups the women also talked about other aspects of their lives and expressed a desire for positive change. For example, the women, because their children were now at school, had become for the first time aware of the concept of literacy. They were also made directly aware of the value of literacy when the discussions about appointing an assistant started. One or two would have had the confidence to undertake the work if they had been able to read and write. Approaches were made to MoSAL who agreed to set up literacy classes, especially for the weaving women, to start in 1985.

One benefit of this was the construction of mixed centres for joint literacy and weaving. In September/October 1984 NCDP from Rustaq had inspected possible sites for literacy cum weaving centres around al-Khabura and the Ministry of Education allocated a teacher. A site in Sirhat was agreed on for a joint centre. Literacy classes began in November five afternoons a week and were attended by 50 women at the start. Heath felt that the centre would also: 'be ideal for group weaving training sessions, and will eventually become the main centre for weaving, for equipment and for buying products and yarn [at al-Khabura].' The weaving part of the centre opened in December 1984. But the idea of having one single centre in the al-Khabura area did not catch on. There were too many constraints on the women's time to allow them to travel to such a centre so, in practice, group training continued in the different villages. Meanwhile in Rustaq the formal training sessions finished in 1985. Thereafter the women worked in their own homes but had continuing contact with the weaving adviser at Ayn Amq as they developed their new roles as independent producers. The women were also exposed to a wider sense of a spinning and weaving group when visits were exchanged between the mountain villages and the coast. In 1985 the women from Ayn Amq made a first formal visit to their colleagues in al-Khabura. It was an invaluable introduction to how they might run their own weaving work from home when their formal training courses were completed. When at al-Khabura 'They talked a lot about money, about products, and about time, and these things have been under discussion ever since' (Heath, 1985b).

Alison Lochhead also innovated with colour and equipment.

The wool produced by many of Oman's sheep is black, which limits its aesthetic potential. However, a smaller number of local sheep have variegated wool colours making them extremely attractive. It is thus quite possible, and was also a traditional activity, to make yarns of different colours, and therefore patterned *mansuls* and other products. White fleeces are also found, and were likely to increase in number through cross-breeding with exotic stock (including the Khabura Project's Chios cross-breds), so the potential for dyeing the wool also existed. In fact, rugs produced in interior Oman had long used dyes, particularly red. Cotton weavers had also dyed yarn and chemical dyes were to be found in the *suqs*. The first experiments with colour were made by Lochhead at Rustaq, using Dylon dyes bought in Britain. Developing a system to handle new dyes was problematic but the women enjoyed working with colour and it allowed their imaginations to flow.

Concerning equipment, the aim was to incorporate ideas developed elsewhere in the world into equipment that could be manufactured and repaired locally, in the hope that this would facilitate the weaving process and allow the production of a greater range of higher-quality goods. The women participated in these efforts because using the traditional ground loom was limiting and tiring; in particular it caused back strain and was very difficult for pregnant women. The engineer on the Khabura Project helped in the process of designing loom modifications (Figure 12). Also, by mid-1985, two floor looms were in use that had been manufactured by the company, Costains, at their workshops in the capital area at a cost of only RO200. Costains also made other smaller precision items of equipment whilst hafs and shuttles were being manufactured in al-Khabura - adding to the pool of artisanal expertise. Some women were again asking about spinning wheels so attempts were made to design a local variant of a wheel then widely used in Turkey and Egypt.

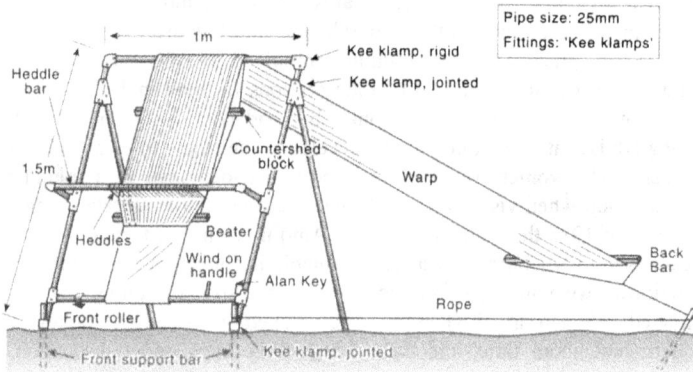

Source: Lochhead, 1984b

Figure 12. Modified ground loom, as tested at al-Khabura

However, in mid-1985 the assessment was that in spite of its problems, the traditional ground loom was still appropriate and that the attempts to improve it had not yielded any result which was significantly appreciated by the women. In Ayn Amq, four women learned to use the floor loom, but found difficulty in mastering its mechanics. But the new tapestry frame was universally successful: cheap, not too big and could be taken to social gatherings. A larger tapestry loom constructed from piping and capable of holding a long coiled warp was attracting interest from five women, who all wanted one. It allowed the weaving of a whole new range of products including larger, finer tapestries.

But the main progress was perhaps conceptual and lay in a growing awareness that bringing about beneficial technical change was possible and that the women could participate in it. Equipment design and development can be an on-going and interactive process and therefore more likely to succeed if it is happening amongst the weaving communities. In addition, such work also helps recreate the concept of mutual self-reliance, in this case between technician and craftsperson, which underlay the rural community develop aims of the Khabura Project as a whole.

It seemed that the women had sufficient technical skills to succeed. In 1986 they had, amongst other important commissions, made three large wall-hangings for Sultan Qaboos University. It was very fitting that the University, Oman's newest and most prestigious institution, should create this link with the country's national craft heritage in a way that was helping to recreate it as a new rural industry.

However, an effective marketing system leading to dependable sales was of critical importance. Alison Lochhead actively involved herself in creating marketing approaches for all the woven products. She tried various means including direct sales from the workshop, shops in the capital area, Dubai and Sharjah, and sales at various company clubs and hotels. She also had many other ideas for selling and sales outlets and stressed that:

'The products must be properly advertised and presented. The image of Omani weaving needs to be heightened to make it an exciting and developing craft and not one that is backward, old-fashioned and only regarded as "bedu" work.' (1984, p. 22)

She saw that there was a resurgence of interest in one particular traditional product, camel trappings. Although camels in northern Oman were now fewer in number and were no longer used as working animals, people, led by the Sultan, were prepared to spend a lot of money on camels for racing and to prepare them in the traditional manner. By contrast she also saw that a high value new market could be developed for tapestries, to decorate the many new hotels and offices that were being built in the capital and elsewhere.

Lochhead, supported wherever possible by the Khabura Project as a whole, made repeated efforts to convince MoSAL of the importance of a marketing policy and structure, but with little result. The essential developmental link between production and marketing did not seem to be understood by NCDP. As she said:

'This lack of marketing policy makes it almost impossible to direct the project in a positive way, resulting in constant dilemmas. Options include: to gear it towards a local market and therefore keep all products strictly traditional, to develop the women's imagination and obvious excitement at their work and gear products more towards an urban and expatriate market, or to look at the possibilities of tapestry production and put the project on a new footing.

Lessons should be learned. Many similar projects throughout the world have failed, not through lack of training or initiative on the part of those involved but through the inability to establish viable market outlets and to gear production to market demand ... Without a marketing policy which works, the whole project will be bound to fail.'

At a meeting in October 1984 MoSAL made some positive suggestions for marketing the products, including an exhibition and opening a shop. In January 1985 a widely publicised exhibition held in Ruwi showed that there was a continuing ready market for the products. Concerning the shop, Heath rightly noted that: 'Ultimately the success of the shop will depend on the quality of the goods it is selling. Therefore it is vital that the manager is able to exercise quality control and refuse products that s/he considers below standard' (Heath, 1985a). The shop opened at the O.K. Centre in 1985 but was closed by the Ministry in November 1986. In practice it had been extremely poorly located with very difficult access. The O.K. Centre mainly attracted shops which sold cheap, imported goods which gave an entirely wrong image to the new weaving products. Much more appropriate and hopeful was a small weaving 'boutique' subsequently rented at the Sabco Centre which was a smaller, very sophisticated shopping mall that attracted people with money to pay for quality goods. From mid-1987 over a six-month period Heath (1987) with others organised a series of promotional spinning and weaving demonstrations at the Sabco Centre, and by this time women from al-Khabura and Rustaq were prepared to participate in selling their crafts in this elegant urban setting.

MoSAL had taken the responsibility of buying directly from the women for resale in the capital area. This showed commitment by the Ministry, and put money into the system, but inhibited the development of a private sector marketing structure and, worse still, did not work efficiently. The woven products in al-Khabura and Rustaq were taken against a receipt but payment was often delayed almost indefinitely. Naturally this affected the morale of the weavers and the quality of their work although a system of rewards and penalties for good and poor work had been introduced for the established weavers, and worked reasonably effectively. The non-payment situation was made worse in 1986 as a result of the collapse of the price of oil. By November, MoSAL owed the women approximately RO3,000 for goods purchased up to April that year. Not surprisingly the number of spinners and weavers had declined to 104 from 220 a year earlier.

Because of MoSAL's very late payments, the project, from about May 1986, had had to be self-supporting through three direct outlets in the capital. This had the advantage of the women receiving payment for their work straight away. It also clearly revealed which products from which weavers were being bought, which had the additional advantage of increasing the shake-out - only the most motivated, responsive and able weavers continuing to work. It was also more realistic. Long-term no small business venture can succeed within the embrace, albeit well-meaning, of a large, bureaucratic, governmental organisation.

One danger of employing dedicated weaving advisers was that the women might become over-dependent on them. For example, it was more socially comfortable for the weavers to obtain yarn from the weaving adviser than to buy it direct from the spinners. When encouraged to do so they complained that the spinners would provide inferior quality yarn, and that in any case there was no way of weighing it. When the adviser provided weighing scales and made it clear that she would no longer be involved in yarn transactions, the women overcame their difficulties (sometimes with the assistance of their children now at school and able to read the scales and record weights and prices) though typically the women only bought from relatives or close neighbours they could trust.

In doing such things for themselves, the women extended to the project their practiced domestic competence as organisers, which had also developed as a result of their menfolk spending a lot time away from home working in the Gulf. Heath (1987) believed that: 'The success of the weaving project can be seen in the skills and organisational abilities that have been developed to the stage where the weavers can achieve independence as a group and move out on their own from under the wing of the Ministry.' The women were keen to take a much more active role in project management. However, an appropriate organisational structure was also needed, perhaps, as Heath discussed, a form of Weavers' Association which could be recognised as a properly constituted small business and therefore obtain training, cheap credit and even the preparation of free feasibility studies and business plans which MCI with the Oman Bank for Agriculture and Fisheries (OBAF) were making available to encourage small business ventures. As a company (or other recognised legal entity):

'They could trade under their own name ... They could advertise and they could have an address and a telephone number so that people and organisations wishing to buy from them or commission work would know how to contact them. They could be visited by other people interested in seeing them at work and in buying things direct from their workshop.' ... 'As a mimimum the only capital they need is for raw materials and for advertising and promoting their organisation, and it is likely that this small sum could be raised from within their own families who were supportive',

as Dutton wrote in a letter to MoSAL in November 1986

MoSAL, however, was not impressed by such ideas, and becoming independent, self-motivated groups of producers may not have been what the women really wanted either. Links with the Ministry offered the possibility, even if illusory, of additional forms of help and protection so, as Heath (1996) suggests, some women will have believed that the more embedded the project became in the Ministry the better.

In summary, in terms of the Khabura Project's community development aims, the spinning and weaving project had created fruitful links between spinners and weavers, between the mountain villages around Rustaq and the coastal villages around al-Khabura, between sheep farmers and spinners and, finally, between equipment manufacturers and weavers. It was also creating in the minds of some of the women, as has been shown, a mental link between practical craft skills and the need for literacy, and extending their organisational abilities. Furthermore, it was widening their world by putting them in direct contact with women's associations, government and commerce. In addition, it was making a fuller use of a locally available natural resource (wool) and earning significant incomes for some women. It was, therefore encouraging mutual self-reliance. It would also have created greater independence from external control if it had succeeded in breaking free from MoSAL - but hidden agendas of the women and MoSAL made the task of weaning the project from the Ministry very difficult. The women wanted MoSAL's (illusory) protection, and for MoSAL the project remained the Ministry's only serious income generating activity involving large numbers of rural women. A sad consequence was that in 1994 (five years after Heath's departure) it was reported that the quality of the work had dropped significantly and that MoSAL's marketing inputs had been reduced to making occasional visits to collect works for displays, taking weeks or even months to pay (Heath, 1996).

4.5 Composting dung

One of the most valuable by-products from a large intensive livestock unit such as the one on the Khabura Project's farm is the manure produced by the stock. It consists of animal droppings, dried forage residues that spill from feeding troughs or racks and straw bedding spread on the floor of the animal pens. Bedding was often used in the winter months when hay stocks were high and the nights cool. General hygiene management required removal of the manure and forage residues from the animal pens. As part of a policy to encourage entrepreneurial activity and to reduce the Project's costs we encouraged customers to buy and cart the manure direct from the pen floors. However, they tended to be selective, taking the droppings but leaving the forage residues in which parasites and bacteria then multiplied.

An alternative approach developed by the Project (Steele, 1983f) was to cart all the manure to above-ground bays with earthen floors and concrete walls on three sides. When the manure was watered the material at the bottom decomposed successfully into compost. However, under the hot Omani conditions, particularly outside the winter months, it was difficult to maintain the top half of the bay moist. The manure dried rapidly and did not decompose. Thus compost pits, dug into the earth, were trialed and demonstrated. A set of four pits 1m wide by 2m deep and 10m long were cut into a small *gelba* near a water supply (Figure 13), with easy vehicle access from the pens and one of the farm gates. Manure produced by the 200 penned animals on the farm at that time filled a single pit in a month. It was layered into the pit so that some 15cm of pure dung interleaved 30-45cm of straw and hay through which oxygen could circulate. The pits were filled in three to four days and watered during filling. For maximum effect 2-3kg of urea was sprinkled onto the straw layers. Once full, each pit was capped with 3-5cm of soil and the whole area flooded and then left. During the winter months little or no watering of the pits was required to keep them moist. In the summer they only needed water once or twice a month, a regime which fitted into the normal irrigation routine for neighbouring *gelbas*.

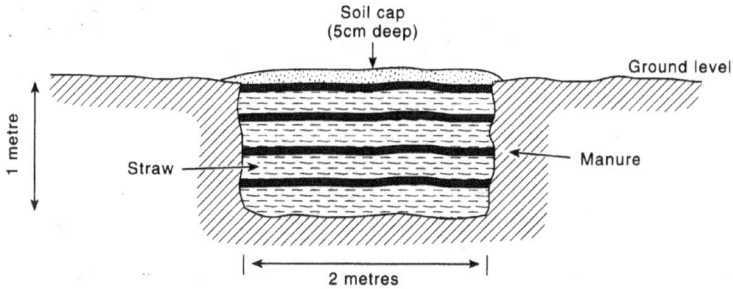

Source: Steele, 1983f

Figure 13. Cross-section of a compost pit

The pits were emptied by hand after three to four months onto the adjoining ground. The compost was then allowed to dry for a few days to reduce weight before bagging and transporting. In practice, the interleaving of dung with straw and hay layers trapped sufficient oxygen in the pits for the complete aerobic breakdown of the animal waste and straw to yield a compost of consistent, good, marketable quality. The main disadvantage of the system lay in extracting the compost from the pits by hand, which was heavy, tiring work. Partial mechanisation would have saved effort and time.

Some of the compost was used on the Project's farm. It was pleasant to handle, easy to apply, and free of weed seed (which had all decomposed). It gave every impression of being an excellent product. Therefore we could envisage a small intensive livestock farmer composting dung in the same manner and using it to maintain the quality of his irrigated forage crop production.

The Project also wanted to know if the product would, in practice, be appreciated not only by local farmers, but also by horticulturists and gardeners further afield. Did the compost have a market value? Could it become the basis of a small but significant rural enterprise which would not only make money but also make full use of a locally available material and be one more step towards rebuilding mutual self-reliance within the village community.

At that period manure (crude manure, not compost) was being imported from Pakistan, and was readily available in Oman at RO1 per bag. In practice this set the price for manure sales of local origin in Oman. The imports could be criticised for undercutting a local enterprise and for importing foreign weed seed and destructive micro-organisms. On the other hand they created a market which had not previously existed, and set a target for quality and price which local entrepreneurs could hope to improve on. In principle our compost, being a superior product, should have been able to command a higher price so long as the market was discriminating. But we did not have the brief or the resources to undertake the research trials to prove the quality either to ourselves or the market. We had to trust to market forces.

We sold the compost by the bag. A wet-filled bag weighed up to 75kg, very difficult to handle and expensive to transport. A dry-filled bag weighed down to 35kg and a locally popular 5-tonne truck could carry 90-100 'dry' bags with ease. We used second-hand bags from flour or animal feed sources. The farm at that time provided 60 or more per month, resulting from its purchases of bagged concentrate feed, and additional second-hand bags were supplied by the Oman Flour Mill (OFM) at 70-200 baisas each. New bags would have been about the same price but a minimum order of 1,000 bags had to be placed. We put thought into the question of building brand loyalty by naming the product with a stencil on inverted bags or sowing on a label (cost: 100 baisas per label), but did not in practice follow through, for want of time.

We also tested the options for sales via a wholesaler or by direct retail. During 1982/3 some 400 bags of composted manure were sold to a company in the capital area, Agricultural Materials Ltd of Ruwi. They bought at the farm gate for RO1 and sold at RO2. For us this was a very simple operation but the gross income per bag was low. We suspected that if we had found the time to negotiate with other companies we would have achieved a higher selling price. Direct sales were made, by medium-term agreement, to the exhibition centre at PDO at RO3 per bag, for 50 bags per month. The bags were transported in vehicles which had to visit the capital for other purposes. As a form of direct advertising, the compost was also gifted to various other individuals in the capital area and to the Sultan of Oman's Air Force (SOAF). It became clear that a large market existed at very good prices, and that for the commercially minded farmer cum businessman the potential was excellent. Other direct sales at lower prices were made to farmers in al-Khabura, at the farm gate.

Unfortunately we were not able to follow through on these very promising leads. Market development was never part of our brief; changes in our own personnel created new project priorities; and there was concern that sales at the farm gate would bring disease into our own livestock. We believed that properly developed, a considerable potential income existed from the sale of compost for those farmers keeping large numbers of goats and sheep under intensive small-farm conditions. Such sales would have had the additional advantage of ensuring that the pens were kept clean. Farmers would have had to decide their personal priorities between own use and sales.

On the basis of our own experience, a herd of 200 sheep and goats gave rise to compostable materials at the rate of 20m^3/month. Internationally compost is sold by the litre, and a large bag holds 80 litres. The total annual production from the sheep and goats would have been approximately:

$$20m^3 * 12 \text{ months} * 1,000 \text{ litres} = 240,000 \text{ litres}$$

$$240,000 \text{ litres/80 litres} = 3,000 \text{ bags}$$

The gross return would have been RO3,900 annually if sold at the farm gate to Agricultural Materials Ltd of Ruwi. For the farmer, this would have been a very welcome additional source of income.

4.6 Soils and tillage

Prior to the Khabura Project the land we had purchased for forage crop production had not been cultivated for as long as anyone could remember - it was simply a stretch of open land adjacent to the coastal date palm gardens. On it was a single *P. cineraria* tree (*ghaf*) and some light *Acacia* scrub.

4.6.1 Soil quality

Evaporation from the land over many years had led to relatively high levels of surface soil salinity which initially impeded seed germination and limited the range of plants that could be grown. However, the farm soils were mainly fine sandy silt loam textures, and free draining. Infiltration rates were moderately rapid so that even with the relatively poor quality of irrigation water used, the soils were suitable for irrigation. Some thin, weak soil capping was evident following irrigation but not sufficient to affect the satisfactory emergence of seedlings (Alexander, 1985).

The soil was not naturally very fertile. The organic carbon content of the soils was low, and showed no appreciable increase even after years of cultivation of Rhodes grass and alfalfa - the combination of high ambient temperatures and soils kept moist by irrigation led to the rapid breakdown of soil organic matter. Total nitrogen levels also remained very low, in spite of repeated use of inorganic nitrogen fertilisers. pH was mainly in the range 7.8-8.2. Cultivation lowered the levels of both phosphorus and potassium. Most critically, when irrigation started in 1976, soluble salts inhibited seed germination. However, the combination of the perhaps excessive volumes of irrigation water used, coupled with the free-draining soil textures allowed rapid leaching of the salts. Thus, for the range of salt tolerant forage crops grown, soil salinity was never a significant problem. Nevertheless, the quality of the irrigation water was always a matter for concern. The sodium absorption ratios (SAR) and the electrical conductivity (EC) of the two principal irrigation wells rose fairly rapidly 1976 to 1980-1 though they then stabilised or improved up to 1984 (Alexander, 1985). By international standards the water was then 'highly saline' and 'medium sodic', but it was still suitable for irrigation given the prevailing soil textures. Analyses in 1988 (Alexander, 1990) indicated an improvement to 'highly saline' and 'low sodic' hazard. In fact, up to 1984 it was shown that the flood irrigation procedure being used had successfully lowered the soluble salt content of the soils, and maintained these lower levels at least to 1988, though there was a slow increase in sodium concentrations in the top soil horizon. By 1988 soil sodium levels had also fallen significantly and there was perhaps a slight continuing fall in EC.

Overall recommendations in 1984 (Alexander, 1985) included measures to increase inorganic nitrogen, phosphorus and potassium, and the use of farmyard manure when available. Also recommended was the addition of some gypsum to the irrigation water in order to counteract the increase in its SAR. By 1988 (Alexander, 1990) the majority of the soils had attained adequate potassium levels but there had been little if any gain in their nitrogen and phosphorus status. A modified regime was therefore recommended.

In 1988 Alexander (1990) also noted some interesting effects of different management practices on the soils. He was able to divide the plots into sub-groups (not entirely mutually exclusive) which he referred to as normal, grazed, ash, compost/ash, nitrogen fixers, gypsum and sprinkler irrigation. The normal group - the control - grew Rhodes grass using standard inorganic fertilisers and was machine harvested. Somewhat surprisingly, perhaps, the grazed *gelbas* showed little difference in any properties that could be linked with grazing. However the ash and compost/ash *gelbas* showed levels of organic carbon, total nitrogen and available phosphorus levels in the top 40cm all considerably higher than the controls. Ash and compost had produced very beneficial effects on the farm soils.

The nitrogen fixers were those *gelbas* in which either alfalfa or Leucaena (both legumes with the potential for fixing nitrogen in the soil) were grown. None of these *gelbas* had received any nitrogen for five months before sampling yet below 20cm the nitrogen fixers showed a significantly higher ($p = >0.001$) level of nitrogen than the controls, and this in spite of the fact that root nodulation was not observed.

Early indications of higher soluble cation levels in the sprinkler irrigation *gelbas*, and other difference from the controls, suggested that it was already possible, after only a few weeks, to detect impacts on soil conditions as a result of the change from flood to sprinkler irrigation. The on-going monitoring of these changes was clearly of vital importance at a time when MAF was beginning to encourage the wide-scale adoption of sprinkler irrigation technologies. Would there be significant net gains in soil quality or crop yields, or otherwise? Regrettably, the closure of the Project's farm in 1989 prevented the continuation of the monitoring programme.

4.6.2 *Rotavator usage, hire and sales*
From the early years of experience at al-Khabura it became clear that problems of land cultivation were the most important single constraint on agricultural productivity in Oman on the land then under cultivation. Because of the lack of

suitable mechanical aids, land was being left out of production or left under a crop that was very low yielding. All too often, as the Durham University surveys had shown, stands of alfalfa were left growing for so many years that each irrigation basin (*gelba*) contained more weeds than alfalfa. The alfalfa was kept because the effort required by the farmers to plough and resow was so great that they preferred to retain the alfalfa until it was almost dead. At that stage the yields were very low and much valuable water was being wasted on growing weeds.

The 4-wheel tractors used by the MAF extension centres in Oman for ploughing were seen to perform effectively on the larger farms. However, on the smaller farms it seemed likely that a smaller, cheaper, 2-wheel tractor might be more appropriate. In order to test this the Khabura Project bought a Howard Gem rotavator which subsequently gave excellent service on both the Project's farm and neighbouring Omani farms (Dutton, 1982a).

The Howard Gem and SuperGem, with 60cm or 76cm rotor width, and with either petrol or diesel engines, cost CIF Muscat RO840 and RO900, respectively, in the late 1970s. The first Gem, purchased in 1976, was initially used only on the Project's 2ha farm. From that time until the project activities ceased in al-Khabura in 1989 all on-farm cultivation was done by Gems and SuperGems to the entire satisfaction of the Project's staff.

By June 1977 the rotavator had aroused sufficient interest among local farmers for one of them to hire it for his garden, at the rate of 600 baisas/hr - the same rate then being charged by the government for the extension centre 4-wheel tractor. In October 1977 the rate for the rotavator was increased to RO1.0/hr, and in October 1978 to RO1.2/hr. Demand for the rotavator was then intense and we could have increased the hire rate but for the 'political' problem of already charging twice the rate required by the extension centre for a tractor that was much more powerful and, seemingly, much more effective. However, seeming was not reality. The Massey-Ferguson-185 then used by the extension centre cost, with a 7-tyne cultivator, RO5,400. Under local conditions on the small Batina farms the tractor worked at a rate of 9hrs/ha whereas the Gem took 12hrs/ha. But for the price of one MF-185 it would have been possible to purchase some 6 Gems. In 12 hours the MF-185 would have covered 1.34ha while six Gems covered 6ha.

Participatory approaches to soliciting from the farmers their comparisons between Gems and MF-185s, revealed the following opinions about the Gems (Dutton, 1982a):

- much more manoeuvrable in confined spaces and small areas;
- work right to the edge of land bordering a cropped area without danger of riding over the crops or of breaking irrigation channels;
- leave a deep, smooth tilth, without clods, and an even surface;
- work easily within large irrigation basins so that neither ridges nor irrigation channels are broken and therefore do not have to be remade;

- work easily on rough and weedy ground and between date palms or other trees;
- using the specially designed 'pictynes' the rotavators penetrate even the cemented Batina soils just as effectively as the MF-185;
- on previously uncultivated land the deep, even tilth facilitates levelling and ridge construction;
- the existence of a larger number of the rotavators would greatly increase the timeliness of sowing and other operations. Many farmers complained that such was the waiting time for the extension centre tractor (up to 6 months) that there were considerable delays in sowing (or the season was lost altogether) with a consequent loss of production.

In addition to the above list of advantages perceived by the farmers, we were also becoming increasingly aware of other advantages of Gems (or other similar pedestrian controlled rotavators) and benefits that would derive from their wider usage:

- their extremely robust construction enabled them to cope with rough handling in difficult conditions;
- because the Gems worked within large *gelbas* they encouraged farmers to enlarge their *gelbas* with consequent significant savings, including: the land area lost to ridges and channels, the numbers of irrigation channels and the time and effort to make ridges and channels. Fewer ridges and channels also meant fewer crop edges, important because it was at the crop edges that weeds concentrated and then invaded the crop;
- because the Gems worked within large *gelbas*, without disturbing the field layout, it was an incentive for farmers to prepare and resow perennial crops. This, for the reasons mentioned above, increased productivity per unit land and per unit water;
- because for the same outlay it was possible to buy six times more Gems than MF-185s, they could have been more evenly distributed within the cultivated area and thus spent a greater proportion of their time working, especially as they could have been carried by donkey cart between distant farms - a cart for this purpose was subsequently designed and built at al-Khabura;
- by the early 1980s a number of local farmers already had experience of operating the Gems for themselves, and the potential therefore existed for the more enterprising of them to offer a contracting service to their neighbours and thereby supplement their incomes.

Because by the early 1980s we had acquired good evidence to indicate that the Gems were not only effective, but also perceived to be so by the farmers, we recommended to MAF a pilot programme for wider testing of them. The elements of the programme (Dutton, 1982a) were:

- the distribution of 10 Gems to farmers in the region of al-Khabura, the initial purchase to be made at a subsidised price or under a loan scheme arranged through MAF or through the subsequently opened OBAF. After purchase, the farmers would bear the full responsibility for the maintenance and repair of their Gems;
- the Khabura Project would act as a co-operative repair and maintenance centre for the 10 machines;
- the Khabura Project would train the farmers in the use of the machines, monitor progress and report to the government of Oman on the value of the impact that the Gems were making on land husbandry and agricultural productivity;
- the manufacturer would train at least one person from the Project in all aspects of maintenance, and supply a full range of spare parts. Because of the need to compete with the almost totally subsidised tractor service, the repair work might be done free of charge to farmers who subscribed to the co-operative but the farmers would purchase the spare parts at cost;
- the Project would report to the government on the value of the machine co-operatives in maintaining the Gems, and also report on the response by the farmers to the co-operative concept.

If we had made the above recommendations in the late 1980s there would not have been the need to propose basing the co-operative workshop on the Project's farm because by then garages were plentiful (servicing the rapidly growing numbers of cars) and had the machinery and technical knowledge which could readily have been adapted to work on rotavators. This also would have made the system more sustainable by being better integrated into the local economy - albeit that the Omani-owned garages were largely staffed by Indian and Pakistani mechanics.

In any case, we failed to convince MAF to finance our proposal and therefore had to stretch our PDO-derived resources as best we could in order to make a start. By 1984 we had purchased six Howard rotavators including two less robust smaller petrol engine machines for on-farm use, and four 11 HP diesel power machines for hire, and the project held a large stock of spare parts. Two of the diesel machines were sold at the end of 1984 and another in 1986, two to local farmers for hire and one to the community of a *falaj* village, Lekleya. Only the large diesel machines were robust enough to be used for such daily hire purposes. At the same time, two full-time Omani operators were being employed and one mechanic was fully trained in small-farm machinery maintenance. The charge rate to farmers in the coastal villages had by then risen to RO1/hr, including the operator's time. The farmer was expected to collect and return the rotavator in his pick-up - most families by then owned a pick-up. The Project continued to hire its own rotavators. Demand for them remained strong, though curtailed by the work being done by the rotavators that we had sold (Table 7).

Table 7. Rotavator hire to small farmers

Item	1982	1983	1984	1985	1986	1987	Total
Hours[1]	300	1,219	1,114	761	768	508	4,670
Farms	95	340	325	201	214	137	

Sources: Dutton, 1986a; Ann. Reps. 1986 and1987

(1) One to three machines available at different times; downturns in 1985 and 1987 related to rotavator sales

The cost to the farmers in the *falaj* villages was higher. Typically the machine remained in the village until all the land had been rotavated, but the farmers had to fetch and return the operator each day which made the work more expensive. However, demand was high and work sometimes continued in a single village for several months (Hillman, 1984b). The very high demand in the *falaj* villages could be explained by the absence of any nearby alternative and to the excellent suitability of rotavators to the confined working conditions between the date palms. In the two *falaj* villages which initially hired the machines, virtually the whole irrigated area was cultivated at least once whereas it had previously been untended for several years. A high proportion of the cultivated area between the palms was subsequently planted to sorghum or grass (grown as shade tolerant forages). Gardens without palms were sown with alfalfa.

In 1983-4 income and costs (including depreciation) of two hire rotavators retained by the Project were fully recorded. In practice they operated for 140 days each per year at a mean of 4hrs/day, earned RO834 and incurred a subsidy of RO598. This was a much better earnings to subsidy ratio than for the extension centre tractors, and would have been better still if the subsidised tractor scheme had not existed. But even in competition with the latter scheme Hillman calculated that it could have become fully, sustainably viable under the following conditions: if the much cheaper but equally robust Daedong 11HP rotavator replaced the Gem; if the operator was also the owner (and therefore had an extra incentive); and if a charge of RO2/hr could be sustained. Under these conditions the annual position is given in Table 8.

Table 8. Projected annual costs and income of operating a rotavator for hire

Item	RO
Operator - 1,000 hours @ RO1.250	1.250.000
Capital depreciation RO300 over 3 years	100.000
Spare parts	300.000
Oil	20.000
Fuel	120.000
Labour for repairs	55.000
Total annual cost	1,845.000
Income @ RO2.000/hr	2,000.000
Gain over salary	155.000

Source: Hillman, 1984b

Thus the owner-driver would have gained RO1,250 income (an attractive salary at the time) plus a small 'profit'. More importantly, the country would have gained a cost-efficient and sustainable system of cultivation of small farms which would have benefited not only the small farmers and the rotavator owners but also a new generation of rural mechanics.

As a result of our work, coupled with some similar findings from other parts of the country, the then American adviser on mechanisation in MAF spent many hours detailing a plan with OBAF whereby, in effect, our earlier proposed pilot scheme (Dutton, 1982a) would have been introduced nationally. However, a new Minister of Agriculture was appointed in 1986 who had an instinctive conviction that large 4-wheel tractors subsidised through the extension centres were all that was required, and the rotavator scheme was stopped in its tracks. The American adviser resigned.

4.7 Forage crops

The technical and economic success of the scheme to rear sheep and goats was in significant measure dependent on the ability of the Project's farm to produce high and sustainable yields of forage crops.

4.7.1 Alfalfa, Rhodes grass and other forage crops

The crop most familiar to local farmers was alfalfa, as already described. We therefore made it the principal forage crop on the project farm, at least initially. It was perennial, and was harvested throughout the year. Growth rates were lowest in late autumn and early winter (Steele and Dutton, 1983a), increased fairly steadily to a peak in June or July and thereafter fell fairly sharply: 'it seems as if the burst of growth in early summer exhausts the crop so that autumn growth, when the weather is still hot, is surprisingly low'. In the late summer alfalfa is prone to weed infestation, the grasses in particular taking advantage of the combination of the heat and weak late summer alfalfa growth rates. In consequence our mean alfalfa yields 1981/2-1982/3 had fallen from 100t/ha green to only 48.4t/ha. On family-operated farms, where family labour was freely available, it was possible to keep alfalfa relatively weed free. However, on small commercial farms, like the Khabura Project's farm, there was no family labour, and neither was there sufficient crop area to justify cost-intensive mechanical and chemical weed control methods. Moreover, in the 1980s family labour was becoming less and less available on any farm because the young men were seeking salaried employment in the Gulf and the children were at school. Even after school hours children were showing signs of reluctance to help in the fields - personal aspirations and the unquestioning sense of responsibility to the family farm were changing. It was therefore difficult to imagine circumstances in which the intensive husbandry skills required to maintain high yields of alfalfa could return. Therefore there was an opportunity to experiment with other forage crops more suited to the new mix of circumstances on the Batina. The Project took up this challenge and in fact must take the responsibility (for better or worse) for introducing Rhodes grass to northern Oman after seeing it growing very successfully in Saudi Arabia (al-Noaim and Farnworth, 1973) in 1976. The proportion of alfalfa on the farm thereafter fell steadily until by March 1983 there were 0.36ha of alfalfa against 1.08ha of Rhodes grass. Growing a forage grass had also become possible, in the local N,P,K-poor soils, because of the introduction and wide availability of inorganic fertilisers.

In practice Rhodes grass proved itself to have the wealth of comparative advantages that had been anticipated (Ch. 3.2.3). It grew densely and competed aggressively with other species, thus minimising weed problems. It proved to be an excellent reclamation crop because some seed germinated on even saline soils, and bare areas were then fully colonised by vegetative reproduction through stolons as successive irrigations began to reduce the salinity levels in the upper soil horizons. But the stolons were easy to cut and remove if this was required so Rhodes did not have the negative tendency of rhizomatous grasses to become an irradicable weed. When mature, Rhodes grass fully demonstrated its tolerance of both salinity and drought - if the irrigation system broke down for a long period the dry grass recovered when the pump was repaired. It also suffered little from diseases or pests. Rhodes, moreover, also responded readily to the summer heat, yielding very heavily in the hottest months of the year. Excess grass produced at this time was easily dried to form hay that was fed to sheep and goats during the winter dip in production. We also grazed part of the farm continually with sheep or goats for years, which saved cutting expenditure and allowed the cost-free return of urine and dung to the land. Rhodes grass also comfortably outlived alfalfa. Thus Rhodes grass fulfilled the requirement of being both very productive and labour saving. It was also relatively easy to grow and because it could be grazed required a minimum of machinery for its cultivation. Its one complexity over alfalfa was its requirement for nitrogen fertilisers.

On the Project's farm year-round records in 1982/3 showed the mean yield of Rhodes grass from the harvested *gelbas* to be 146 t/ha (green), some three times the alfalfa yield over the same period. In practice each *gelba* was harvested about six times in the summer half year, and only 2-3 times in winter, reflecting slower growth in the cooler season (Steele and Dutton, 1983b).

On our small farm we experimented with different cultivation practices. For example, seed rates of 0.25kg to 1.5kg per *gelba* of 200m^2 were tried, but once the grass was established there was no apparent visual or yield difference apart from the fact that the higher seed rates gave an initial much higher population of grass (and more defence against weeds when Rhodes grass followed alfalfa). A common seed rate used by the Project in the autumn was 0.75kg per 200m^2, equivalent to 38kg/ha. In spring sowings, with hot weather to follow and less stolon growth to colonise bare patches, the seed rate was increased to 1 kg/*gelba* or 50kg/ha (Steele, 1983e). Our general aim was to devise a system which required minimal initial outlay and minimum maintenance.

Wherever possible we recycled crop residues and animal waste - converted to ash or composted, as already described - back into the forage crops, but regular applications of N, P and K were also essential, as successive crop and soil analyses revealed. We followed Al Noaim and Farnworth (1973) and Farnworth and Ruxton (1974) who, in trials at Hofuf, Saudi Arabia, found that dry matter yield from Rhodes grass increased proportionally as nitrogen application increased from 200 to 500kg/ha/yr. When the detailed soil analyses in 1984 (Alexander, 1985) revealed shortage of all major nutrients, application rates of nitrogen fertiliser were increased to 300kg N/ha/yr, while applications of phosphorus and potassium were increased to an equivalent of 150kg/ha/yr each. Two-thirds of these rates were advised for the grazed *gelbas*. But in 1988 it was observed that only minor, if any, improvement in soil nitrogen and phosphorus status had occurred so the recommended rates were again revised upwards. There were also imbalances in the status of minor nutrients of which perhaps the most important was shortage of copper which is associated with the disease known as swayback in which goat kids lose control of their legs, leading to death.

As already noted, we experimented with grazing sheep and goats from the outset. A third of a hectare adjacent to a line of goat and sheep pens was cultivated as eight long, narrow *gelbas* and sown with Rhodes grass in 1977/8 (Figure 14). The *gelbas* were grazed permanently thereafter until the project farm closed in 1989, with the obvious primary advantage of never having to face the costs of mechanical harvesting and stall feeding, or re-cultivation. The *gelbas* were isolated by electric fencing and grazed - at times - in rotation. For the most part, the sward remained dense and weed free, and therefore entailed very limited maintenance costs. Under these conditions goats were better grazers than sheep. The sheep tended to graze some shoots to ground level while leaving other stems untouched because for the sheep they had grown too fibrous. The goats tended to maintain a much more even carpet. Grazing the sheep and goats together was also successful, though difficult to organise because of the need to return them to their separate pens in the evening. The initial success with grazing encouraged us to extend the experiment in later years, even constructing new pens adjacent to strips of Rhodes grass in other parts of the farm (Figure 14). From an experimental point of view grazing unfortunately prevented measurement of yield. However, visually the grazed *gelbas* looked vigorous and productive though if the total area was too large for the number of animals using it the grazing became patchy, particularly where sheep were concerned. The variability of Rhodes grass yields from season to season made this a significant management problem.

To harvest both Rhodes grass and alfalfa we imported reciprocating blade, pedestrian controlled, powered cutters, initially the Allen Scythe and Howard 'Dragon'. These machines, driven by local operators, worked reasonably successfully within the large *gelbas* into which we divided the field. Because they were potentially more dangerous than the rotavators in inexperienced hands we never made serious attempts to introduce them off-farm, and in practice the local small farmers never showed any interest in them, preferring to continue harvesting with the traditional sickle.

Weeds were a problem, particularly in alfalfa, as has been noted. The reciprocating blade cutters harvested the alfalfa 3-5cm above ground level. Grasses below this level were not removed, as they would have been, by the women, in gardens where the traditional sickle was used. Conversely, they were actually encouraged especially in late summer when they grew much more vigorously than the alfalfa crop. We believed that manual labour would become scarcer and more expensive in future years, when families were better educated and with a higher proportion of wage earners, so that hand cutting and weeding would inevitably decrease. Therefore mechanical cutting of forage crops would increase, and solutions would have to be found to control the weeds. Herbicides were an option, but costly and with their own dangers. We experimented with rotating between alfalfa and Rhodes grass with the hoped for benefits of disrupting disease and pest cycles, smothering weed grasses and improving total yields (Steele, 1983d,e). In practice Rhodes after alfalfa did smother low-growing weeds because of its rapid growth in summer, and production of stolons in winter. Rhodes also responded much more strongly than the weed grasses to nitrogen fertilisers, which additionally checked weed growth. Nevertheless, to ensure that the new Rhodes grass started successfully it was found to be advisable to rotavate the ground several times following the removal of alfalfa so that successive crops of germinating weeds could be eliminated. In practice the rotation cycle tended to end with the change to Rhodes grass - thereafter we could see no reason for reverting to alfalfa again!

From time to time the crops were analysed for nutrient value, as summarised in Table 9 (Dutton and Steele, 1983). Nutritionally there were some important differences between samples, depending on age and condition. For Rhodes, the variation in crude protein levels between the leafy samples with 14-16 days growth (mean 13.4%) and the mature headed samples (mean 9.1%) is entirely explicable in terms of age of crop before cutting. It illustrates the importance of cutting the crop young. The results from the grazed *gelbas* indicate that grazing achieves effectively the same result.

For alfalfa, it seems (if one ignores the somewhat doubtful analysis made in Durham) that the fully mature crop contains about 16-17% crude protein and the younger cut crop 19-30% crude protein (mean 22.4%). As a sheep and goat feed this is probably unnecessarily rich in protein - good quality Rhodes grass is sufficient.

Estimations of metabolisable energy (ME) indicate that the stemmy mature Rhodes sample, with high fibre, had a low energy content but that younger samples had very acceptable energy levels. Fibre content was lower in young Rhodes, and in legumes.

Additional analyses of Rhodes grass undertaken on 10 samples in the years 1983-9 (Barker, 1989) showed crude protein levels at 10.1-15.3% after 2-3 weeks growth, but only 7.1-8.6% after five weeks growth. Metabolisable energy and fibre levels were much as before. To some extent the preferred date of cutting depended on circumstances. Less frequent cutting is less costly and provides more bulk, and dietary deficiencies can be made good with other feed concentrates. More frequent cutting yields less bulk and is more costly but needs fewer feed supplements - and is more appropriate if animal stocking rates are lower. Other locally grown and purchased feedstuffs, including dates, were analysed for their feed value, enabling balanced diets based on locally grown or otherwise available feeds to be prepared and used cost effectively.

Table 9. Crude protein content of different crops grown on the project farm

	Year	Age/condition	Year sown	Lab.	Crude protein	ME MJ/kg DM[1]	Fibre MAD g/kg[2]
Rhodes	1978	(young)	1977	Oman	15.3		
	July 78	14-day growth	1977	Oman	13.6		
	July 78	16-day growth	1977	Oman	12.3		
	1978	14-day growth	1977	Oman	13.0		
	Dec 82	Mature (+ heads)	1982	Durham	8.6		
	Dec 82	Mature (+ heads)	1982	ADAS[3]	9.6	7.76	392
	Apr 83	24-day growth	1981	ADAS	20.4	9.6	335
	Apr 83	10cm high - being grazed	1978	ADAS	12.8	8.6	358
Rhodes + siratro	Apr 83	10cm high - being grazed	1982	ADAS	16.4	9.0	345
Alfalfa	1978	(young)	1977	Oman	26.9		
	May 78	?	1977	Oman	19.1		
	July 78	27-day growth	1977	Oman	16.5		
Alfalfa hay	Mar 78	?	1977	Oman	21.5		
Alfalfa	Dec 82	stemmy		Durham	22.3		
	Dec 82	stemmy		ADAS	17.0	288	9.5
	Apr 83	17-day growth		ADAS	22.1	288	9.9
Siratro	Dec 82	stemmy	1982	Durham	18.1		
	Dec 82	stemmy	1982	ADAS	16.6	308	9.2
	Apr 83	less stemmy	1982	ADAS	21.7	289	9.3
Clittoria	Dec 82	stemmy	1982	Durham	16.9		
	Dec 82	stemmy	1982	ADAS	15.4	235	10.4

Source: Dutton and Steele, 1983

(1) ME MJ/kgDM = Metabolisable Energy, Mega Jules/kg dry matter
(2) MAD fibre = Modified Acid Detergent fibre; high fibre is poor nutritionally
(3) ADAS = Agricultural Development and Advisery Service

The project also experimented with other forage crops, notably the legumes siratro, stylo, lablab and leuceana (Steele, 1983d). Soil analyses (Alexander, 1990) indicated that they were fixing nitrogen. However, the potential advantage of this in mixed siratro/Rhodes grass plots was lost because the grass grew too vigorously for the siratro, which was smothered. Leucaena in practice was mainly restricted to one border of the project farm where it acted as a break against dust from the adjacent dirt road.

Other analyses showed that the copper content of the alfalfa was low to lowish, and that of the Rhodes grass rather low at 7.3-9.4ppm. However, interpreting the significance of these levels for swayback in goats is difficult because its true availability is also affected by other minerals, even minerals breathed by the animals in windborne dust.

4.7.2 Working with small farmers

When nearby Omani farmers who regularly visited the Project's farm saw Rhodes grass growing successfully they expressed an increasing interest in growing some on their own land. In discussion with the farmers, it became clear that they were particularly impressed by its high yields, perennial growth, absence of weeds, palatability for all forms of livestock, and good grazing potential. In the spring of 1978, therefore, small patches of Rhodes grass were grown, under careful supervision on two Omani farms. Integral to these initial trials, basic fertiliser experiments were conducted on the two farms, and on a few others in the locality. We had to be sure that the farmers understood that unlike alfalfa, Rhodes grass would not produce a worthwhile yield without nitrogen fertiliser. Initial experiments were deliberately simple - adjacent *gelbas* of the crop were given some nitrogen (ammonium sulphate or nitrate), or left without. In practice the *gelbas* with nitrogen were so clearly and visibly more productive than those without that the need to use nitrogen fertilisers was rapidly understood.

Some farmers fed the grass to their animals, and others grew it for sale in Al Ayn where the price was not much less than that for alfalfa. However, for these farmers the wet winters of 1981/2 and 1982/3, coupled with the high levels of production of Rhodes grass for sale at the large, internationally financed, commercial farm at Sohar, known as Oman Sun Farms, made the production of Rhodes grass less profitable because of the fall in prices in Al Ayn. Therefore some of them stopped production, or left their grass to dry, without irrigation. When the price later started to rise again, they subsequently re-irrigated and discovered the grass's remarkable powers of recovery Other farmers grew alfalfa for sale and produced Rhodes grass to feed to their own stock. In general these people wanted to increase their stock numbers and therefore to expand their Rhodes grass area.

Because of the drop in Rhodes sale price and opportunities in 1982/3, the farmers bought less seed from us (Table 10). However, in the second half of 1983 sales picked up again as the 1983/4 drought reduced the availability of natural grazing and browse. From 1984, sales remained at over 200kg each year to 1987 (when we stopped keeping records) by which time the crop had become fully established locally. The farmers' interest in growing Rhodes for sale, and having therefore to market a product of good quality, created the opportunity to teach the small farmers a lot about methods of cultivation. But the Project's hope was that as farmers' stock numbers increased, a higher proportion of their grass would be fed to their own animals.

Farmer levels of expertise varied considerably. Some farmers reported on by Barnwell (1984b) had been growing Rhodes for five years extremely competently, with annual yields similar to or even greater than yields on the Project's farm. The largest areas were up to half a hectare, though about 400m^2 was much more typical. The farms were situated up to 10km either side of al-Khabura village, in a total of over 30 separate village communities by 1987.

Table 10. Sales of Rhodes grass seed to local farmers 1981-7

	1981	1982	1983	1984	1985	1986	1987
Seed sold (kg)	103.9	121.8	70.1	198.5	225.0	204.0	216.5
Farmers buying seed (nos)	105	147	91	165	179	168	140
Villages where seed was sown (nos)	20	26	24	27	30	29	31
Area sown (ha)[1]	4.2	4.9	2.8	7.9	9.0	8.2	8.7

Sources: Barnwell, 1984b; Dutton, 1986a; Ann. Reps. 1986 and 1987

(1) At rate of 25kg/ha

Perhaps our greatest extension contribution had been to demonstrate that Rhodes grass could be grown successfully between the widely spaced date palms that characterise the palm planting regimes on the Batina coast. There was sufficient light for the grass to grow well, and Rhodes was also tolerant of the saline soils and water that were typically found in the older palm gardens on the coastal side of the new main road. The crop gave a new lease of life to many gardens and even, in some cases, rejuvenated the palms themselves which benefited from the additional irrigation water being applied. Some farmers also experimented with housing their sheep and goats within the gardens in ways which gave them easy access to the grass. The animals also benefited from the more shady conditions, and were able to supplement their diet with dates in their season because the poor quality coastal dates could no longer find a market for human consumption (Ch. 2.2.3) now that a much wider range of foodstuffs was available even in small rural markets, thanks to oil-funded food imports.

As part of the process of developing skills and entrepreneurial activity within the local community, we placed the responsibility for the sale of Rhodes grass seed into the care of a man from al-Khabura, a man who was also a Project employee. As he was an employee it remained possible to monitor the quantity of seed sold and the number of farmers who made purchases. But by the end of 1987 it was noticed that other local merchants were also selling Rhodes seed as a direct result of the increased interest in and knowledge of Rhodes grass acquired by the farming population as one outcome of Project activity in previous years. During the 1980s Rhodes grass had also become the forage crop favoured by the commercial farm sector, with whom we were not involved. They too had come to appreciate its manifest strengths, and they were able to sell the grass, as bailed hay, into the Emirates and, increasingly, into Dhofar where the market to feed the mountain cattle seemed to be inexhaustible. The downside of this development was an excessive use of water which was, in effect, being exported from the Batina. Attempts to control this commercial scale activity have been a feature of the 1990s as the Ministry of Water Resources has sought to increase public awareness that water is a finite resource and must be used wisely.

4.8 Irrigation systems

Irrigation systems assumed an ever greater importance in the Project's range of activities even though the farm system under development had goats and sheep as its primary focus. All agriculture in Oman requires irrigation because the rainfall is low and very irregular. The Project, therefore, had to irrigate its alfalfa and Rhodes grass which created the opportunity for experimentation and change throughout the years of our development work at al-Khabura, both on the Project's farm and off-farm in gardens belonging to our neighbours. Gradually therefore, irrigation grew in status within our priorities until it equalled the importance of the sheep and goat work. In partnership with the farmers we tested, developed and demonstrated various approaches to irrigation. Progress on them all is discussed in this section with the exception of the work on the fibre-reinforced-cement channel liners. They are considered separately because they are also an example of an attempt to create a small-scale local manufacturing enterprise.

4.8.1 Modifying flood irrigation

The Project undertook a lot of work on flood irrigation, though not because it had a policy to advocate flood irrigation above all other means of taking water to the plants. It was simply the method of irrigation with which local farmers were familiar, and the Project sought to modify and improve the technique by enlarging *gelba* size and by changing the channel design and layout. The Project farm's *gelbas* were flood irrigated in rotation. The application rate and frequency initially, and deliberately, followed local custom and practice. In general, this meant filling a *gelba* to nearly the height of a ridge and irrigating very approximately once a week. Irrigation rates were not measured and although in subsequent years various attempts were made to monitor them it was not until May 1987 that an irrigation schedule was introduced. By then water meters of good quality had been purchased, and sufficient meteorological and soil data had been collected on which to base an irrigation frequency and application rate to cater for crop requirements, evaporation and leaching. Day to day operation of the schedule remained, however, in the hands of the same uneducated operative as before and he tended to ignore the schedule and maintain his original rhythm of irrigation. But at least the volume of water used each day was recorded, as were the *gelbas* irrigated, for a continuous period from June 1987 to end-May 1989, when the Khabura station closed.

The data were computer-logged and analysed to give the depth of water in each *gelba* at each application. Monthly and annual totals for each *gelba*, or group of *gelbas*, could then be totalled, and the annual running mean depth of water that had been applied to each *gelba* calculated. These running means started high, at 5.9m depth in the year from June 1987, and rose to 8.4m depth in the year from January 1988 from when they remained fairly constant. These were very high volumes of water, 8m depth being equivalent to 80,000m^3/ha, even though in reality the long-term average may have been nearer six than eight metres - there is statistically clear evidence that the operator of the flood system increased his application rate when the opening of the sprinkler network (Ch. 4.8.2) on part of the field meant that he had fewer *gelbas* to irrigate. But 6m depth, which corresponded closely with the application rate previously calculated for the local farmers (Letts, 1982), is about double the theoretically required rate as estimated from a modified Blaney-Criddle formula. But whether theory or long-established local practice is correct to cover the requirements of crop growth, evaporation, transpiration and leaching is a bigger question which requires more detailed examination (partially accomplished by later Project work, see Ch. 5.3.1-2) before sprinkler systems for forage production are generalised, particularly where there is a danger of surface salt accumulation as a result of inadequate leaching. Certainly, however, observation of the Project's irrigator at work did not encourage the belief that local practice is necessarily intuitively correct. Quite apart from increasing the annual depth to 8m when more water became available, different strips of *gelbas* received significantly different annual totals and monthly distributions of water (Dutton, 1991).

Following local tradition and current practice, the Project decided to lift the water required to irrigate its forage crops from wells hand-dug on the Project's farm. The first of these, dug by the building contractors primarily to meet the needs of the housing and offices proved to be too shallow and too small in diameter to cope with the water demands of the forage crops and the livestock. It was also carefully lined with concrete rings (not, in fact, necessary on the Project's farm where the shallow depth to water and the densely packed subsoil made erosion and collapse of well sides unlikely) which made it difficult and costly to increase well diameter or depth. As the well was near the houses it was also downstream and down slope of the main forage area which meant that the water had to be pumped to the top of the field before it could flow under gravity onto the crops, and it also meant that when excess, leaching, water carried salts back down to the water table they fouled the well, whose soluble salt content rose. The use of this well therefore was initially restricted to domestic supply but when the salt content rose to the point where the water tasted brackish, domestic water was drawn from one of the new wells, described below, though eventually, for drinking purposes, bottled water was purchased for all staff.

119

For irrigation, the so-called 'road well' (Figure 14) was dug in 1977 to the side of the forage area adjacent to the main graded road leading into al-Khabura village (Figure 4). One reason for siting it there was easier access from the generator for a cable to supply power to the electric submersible pump which we used. This pump was an innovation made in the subsequently justified belief that the electric submersible would require much less servicing than the 6.5HP Petter diesel pumps which had become the standard replacement for the traditional *zagira*. A three-inch pipe carried water to the high inland side of the forage field. Water then travelled along the top of the field in a horizontal channel and finally down secondary channels under gravity to individual *gelbas*. Another, short, pipe from the 'road well' fed directly into a channel bordering the eight long, narrow *gelbas* of the original grazing area. Later, in 1981, when the field was under full crop cover, it became apparent that one well could not supply the whole field so the so-called 'new well' (Figure 14) was sunk beside the long horizontal channel and also pumped water into it where it mixed with the water from the 'road well'. The 'new well', which had the best quality water, also took over the role of domestic water supplier.

Originally the channels themselves exactly followed tradition. Their sides were shaped, with the assistance of the *graz*, by heaping soil from the field and compacting it by foot to make it watertight. Gates were cut away to allow water to flow from the secondary channels into the *gelbas*. Throughout the channel system water was diverted by constructing or removing temporary earth dams.

However, although the earthen channel network was low cost to construct, in practice it proved to be both inefficient and high cost to maintain. It was also high cost to operate as it required the near continual services of one man when the pumps were on. As the Project had moved away from being a family farm to being a small, albeit trial, enterprise without access to family labour, the costs of operating and maintenance assumed greater importance. In any case, local labour was becoming scarce because of the growing drift of young men into the Gulf to find work, as noted above, so labour saving was becoming more important on the family farms also. A main cause of the inefficiency of the earthen channel was that a relatively large volume of water was used to wet the channels and then evaporated unproductively from them. Our previous survey work had made us aware of this problem even on the typically small patches of alfalfa then grown by most farmers. However, our larger farm area and greater length of channel made this inefficiency much more manifest, even though the Project's larger *gelbas* (20m by 10m) greatly reduced the total number of channels per unit area. It took a long time for the water to reach the *gelbas* and then the rate of delivery was too low to permit an even distribution of water across the surface of the enlarged *gelbas* which we used in order to increase the proportion of land under crop and to facilitate the access of machinery.

So, in an attempt to reduce inefficiency and running costs the Project started to experiment with channel liners. First, we experimented with plastic sheeting which had by then appeared on the local market, though not for this purpose. However, most of the available sheeting was transparent (allowing weed growth underneath it), none was resistant to ultra-violet light, and none was of the dimensions we required. We therefore obtained from Britain long roles of 2m-wide black, ultra-violet light resistant polythene sheeting and experimented with that. The sheeting was fairly easy to lay if the start was tucked into the earth around the upper end of the channel while the role straddled the channel and if, as the role was unwound, water was poured onto the sheet to hold it firm and to shape it onto the earthen channel bed. Then the edges of the sheet were successfully tucked into the sides of the channel. The use of the sheeting meant that water had to be siphoned from primary to secondary channels and from the channels into the individual *gelbas*. To ensure sufficient depth of water at the points where the siphons sucked from the sloping secondary channels we designed a mobile dam. A canvas skirt was stapled to a semi-circular plywood shape, attached to a wooden bar stretched across the channel, such that the weight of water pressed the shape and the skirt against the channel wall to make a nearly watertight dam which backed up a pool of water several metres in length. It was possible to irrigate a single *gelba* by running up to three 2" plastic pipe siphons from the pool, each separated by 1-2m from the others. The separation ensured a much more even distribution of water across the *gelba* surface, whilst the use of three small-diameter pipes reduced the rate of flow per pipe and therefore minimised erosion at the point of exit within the *gelba*. Siphons also tend to be self-correcting; that is to say if the water level in the source is being drawn down too quickly, the head is reduced and the rate of flow slows, and vice-versa. In consequence, it was possible, with reasonable confidence, to leave the siphons to operate unattended, only returning to move the dam and the siphons to the next *gelba* when the first had received its quota of water.

The use of the sheeting and siphons also clearly demonstrated other benefits. No water was required to wet the channels and so it moved through the channel network distinctly faster with no loss of water. The system, therefore, achieved a much more rapid and even distribution of water across the *gelbas*. The black, light-impermeable sheeting prevented any growth of weeds and stopped evaporation losses from the earth channels (which were kept permanently moist by lateral movement of water through the soil). The replacement of gates by siphons also eliminated the need for gates and thus the erosion of channel walls normally associated with them.

But some problems remained and others became apparent. In practice the width of the channels, crop edge to crop edge, was almost two metres on average, representing a very considerable loss of land within the field. Covering the channels with polythene had done nothing to reduce this loss of land. An unanticipated problem was caused by the fact that after irrigation some water remained in both the primary and secondary channels. This, in Oman's hot, arid climate, proved to be attractive to birds and in particular to the Indian hooded crows. As they perched on the sides of the channels and drank from the pools, they holed the polythene with both claws and beak. Not only was water lost through the holes but the channel beds softened and occasionally disintegrated. The beds were then very difficult to repair and it proved almost impossible to recompact them without removing the polythene liner altogether. When the holes in the polythene were noticed early enough it was possible to repair them with the use of a polythene cement and patches, but for a patch to remain watertight the polythene had to be completely dry and free from dirt - very difficult under the circumstances. Also the materials, skills, costs and time needed for all these repair jobs negated an important advantage the polythene liner was supposed to have over the traditional earth channel. Another problem was learning the knack required to start a siphon which the relatively wide, shallow channels made more difficult to master. But the worst problem of all was that the polythene, in practice, was not very resistant to ultra-violet light. Within a year the material perished and started to tear. Wind got into the tears, extended them and sometimes lifted long lengths of sheeting from the channels. It became clear that an alternative channel liner had to be found. Our experience with these is discussed in Ch. 4.9.

4.8.2 Introducing sprinklers

It was decided to convert part only of the Project's farm to sprinkler irrigation in 1987 not only so that it could be compared with the flood irrigated area but also because it continued to seem likely that flood irrigation would remain the system used by the majority of small farmers for many years to come (Figure 14). The potential advantages of sprinkler irrigation were assumed to be: lower absolute water usage per unit area; lower water usage per unit of forage production; greater control over water usage, in order to match supply with changing seasonal demand; a higher yield per unit area because no land would be lost in ridges and channels; and, because there were no ridges or channels, no time spent making or repairing them, no weed or salt-burn problems associated with them, and easy access for cultivation and harvesting machinery. But set against these potential advantages were some definite and potential disadvantages: higher capital cost, including well-head equipment and lateral pipes; the need for a greater understanding of hydraulics, and of crop water requirements, if sprinklers were to be used efficiently; risk of soil salination if insufficient water was used to leach soluble salts; increased evaporation losses; and heat scorch if water with high salt and/or high carbonate content was used.

It was hoped, on the Project's farm, to test the potential advantages and problems of sprinkler irrigation over an extended period but by the time the sprinklers were fully operational, irrigating newly sown Rhodes grass in May 1988, only 13 months remained of the Project's life, to end-May 1989. Therefore, although some results were obtained, they were partial and the conclusions tentative. However, they are interesting.

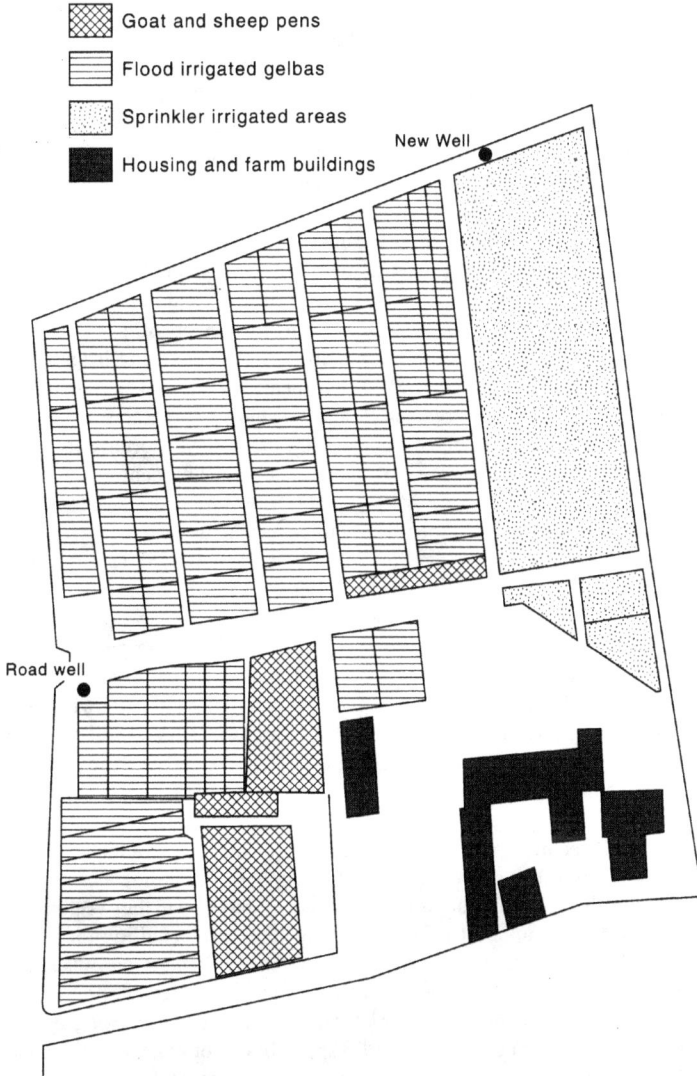

Source: Dutton, 1991

Figure 14. Layout of the Khabura Project's farm, 1988/9

The equipment and material costs for the 0.35ha converted to sprinklers, totalled RO1,752 including an electric submersible pump (RO582) and a fertigation unit

(RO380) which would have sufficed for a larger area. Costs per hectare were in the order of RO3,160, excluding labour or transport of equipment from Muscat.

Compared with the previous *gelbas* the non-cropped land was reduced from 16.5% to only 2.9% of the total area, representing a significant increase in crop area. Also, for the seven months June-December of 1987 and 1988 for which a direct comparison could be made, the sprinklers used 41% less water and produced an annually rated yield of 126t/ha fresh Rhodes grass. The Omani farm staff were favourably impressed with the saving of labour and time in field preparation, and in cutting and carting the crop (Dutton, 1991).

However, as stated above, these results and impressions of the benefits of sprinkler irrigation could only be tentative considering the few data on which they were based. They would only be of value if they could be confirmed for a much longer time series and for data from a number of sites throughout Northern Oman. Such data were potentially available from the irrigation units installed on the farms in the Goat Multiplication and Development Project (GMDP) (Ch. 5.3.1). We were keen to make a study of them to see whether, in practice, they increased or decreased water use efficiency. Fortunately we were given the opportunity when the UK's Overseas Development Administration (ODA[14]) seconded an Associate Professional Officer (APO) to the Project for 18 months from December 1991 specifically to undertake this work.

At government level during the late 1980s increasing awareness of and concern about the huge and growing volumes of water being inefficiently used by the agricultural sector led to some action. The introduction of sprinklers to GMDP was one example of the action taken by MAF. More generally, MAF, convinced that action was urgently needed and having to think nationally, took the route of working through local agents of internationally known irrigation companies. At the same time it asked FAO to prepare detailed soil maps of the Batina farms with the intention of making exact matches between soil capability, crop choice and irrigation technology. This was, of course, a perfectly rational approach to obtaining a general improvement in the national level of water use efficiency and seemed to combine speed with scientifically validated solutions. Nevertheless, questions remained about the degree to which uneducated farmers would apply the answers suggested by science, and about the wisdom of the government taking upon itself so much of the responsibility for effecting fundamental changes to age-old irrigation systems. Farmers are businessmen, and above all they need to make a profit which means that scientific niceties are only one of the groups of factors which affect their decision-making processes. It was important to leave the feeling of responsibility for change in the hands of the farmers just as they, without significant government intervention, had already replaced all their *zagiras* with diesel pumps and therefore without question took full responsibility for their servicing, repair and replacement. If the government had been prepared to listen we would have argued, therefore,

---

[14] Renamed Department for International Development (DFID) in May 1997.

that the encouragement of small, local, private 'Irrigation Services' companies, given proper training but using technologies which they could readily manipulate and mould themselves, would better retain responsibility in local hands and be more responsive to the farmers' ever changing needs. We would also have argued that although change based on such an approach would have started more slowly, it could have gathered momentum more quickly and produced better results in the longer-term, including those community bonds created by the mutual self-reliance of its people.

4.8.3 Trickle irrigation, including small-farm trials

Originally irrigation was only of concern to the Project as an element of forage crop production, itself one element of the small-farm, small ruminant livestock management system under development. It was only when the growing local awareness of water shortage encouraged the Project to elevate the status of its irrigation interests that, in 1983/4, attention was given to the irrigation of non-forage crops. The result was the development of a low cost, low pressure and low technology trickle irrigation system (Stern, 1979) suited to row crops and fruit trees (Hillman, 1984a, 1986a). The design criteria of the system were as follows:

- that all or most of the elements of the system should be cheaply available on the local market;
- that the system should be low head, and therefore capable of being operated by farmers with the typical low pressure pumps already installed in all farmers' wells;
- that it should reliably deliver sufficient water to the crops to ensure high and sustained yields without undue water loss;
- that it should be capable of assembly, installation, modification, maintenance and expansion by local and locally trained Omani technicians, who might form themselves into small companies to undertake the work.

The soils of the area were sandy loams, and therefore suited to trickle irrigation in that they would readily transmit water both vertically and horizontally. The water itself was free from physical contaminants but was high in carbonates and had variable to highish soluble salt concentrations (range from 1,000 to 5,000 micromhos). Carbonates can clog emitters, and high levels of soluble salts can cause soil salinity when water application rates are restricted by trickle or drip systems.

The 6.5HP diesel pumps used by most farmers could deliver water at the rate of 10 litres/sec at the well head, and lift water into an overhead tank. But they could not, for example, drive a commercial trickle system requiring 15psi or 10.5m head. A low pressure system would avoid capital expenditure on a new pump and the high cost of pressure-rated piping. Non pressure-rated pipe was available on the local market, both rigid PVC which could be used for mains and sub-mains and flexible alkathene for the lateral pipes to deliver the water to the row or to the vicinity of the individual tree.

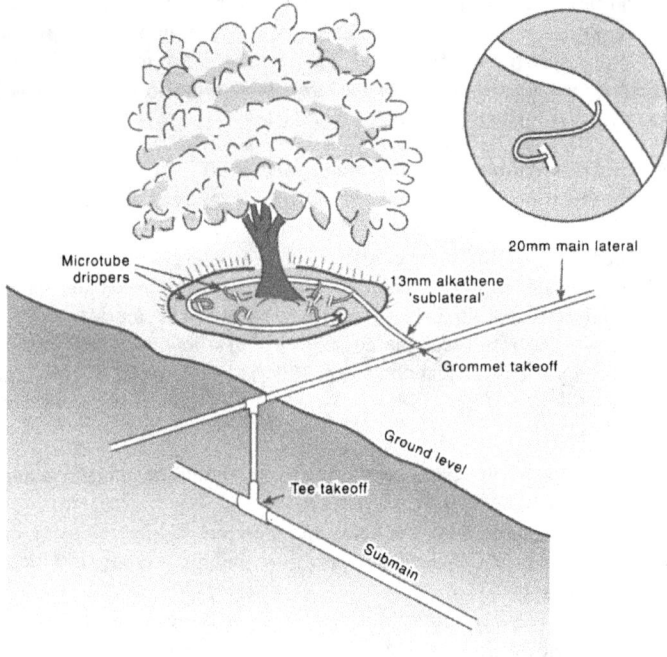

Source: Hillman, 1986a

Figure 15. Schema of low-head microtube dripper installation

The emitters themselves consisted of cut lengths of large bore microtube (1.27mm diameter), imported in 200m rolls. One end of the emitter was pushed into a hole in the alkathene pipe made with a simple hand tool (which invariably made a firm, watertight joint). The emitters were sufficiently wide bore to minimise blockage due to the deposition of carbonates. In practice blockages that did occur were easily cleared by lifting and flicking the end of the microtube. Successful operation also depended on a regular, though very infrequent, check of all tubules to ensure that none was clogged with earth or weed roots. We came across a few instances of tubules where roots had penetrated so far and so tightly up them that the tubule had to be removed and replaced. But this cheap and simple operation was easily within the capacity of the farmer to perform. In fact, of course, for farmers who managed their crops properly it was very rarely necessary. The emitters could also be cut to any length as an easy method of controlling the rate of flow; long emitters had high total internal friction and therefore a low flow rate. If the flow rate was too high, because the emitter was too short, it could cheaply be reduced by replacing the emitter with a longer one.

In the case of vegetable row crops emitters were inserted into a lateral pipe equidistantly, sufficient to ensure permanently moist soil around the root zone. In the case of tree crops, a sub-lateral line encircled the tree, arranged so that a cluster of emitters distributed water widely over the total growth area of the roots (Figure 15). As the roots grew, the sub-lateral could be spread more widely and extra emitters added in order to deliver more water. But of course the system when first installed had to be sized to allow for this increase in water demand. The end of each sub-lateral was folded over and held with a removable ring which could be opened, if ever required, to flush the pipe, to leach salts or to rot manure placed around the tree.

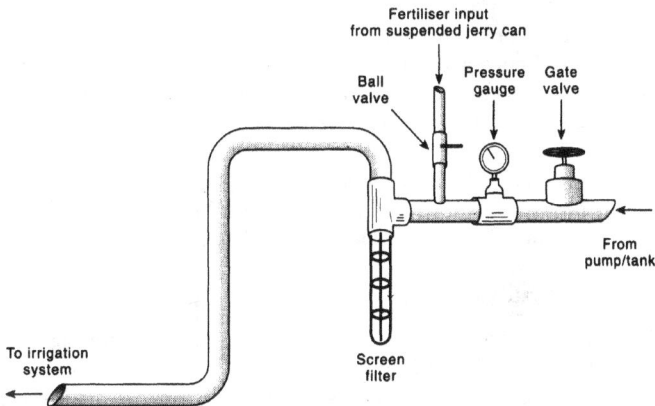

Source: Hillman, 1986a

Figure 16. Schema of typical headworks for the low-head microtube dripper installation

One problem to overcome was balancing the desired tubule delivery rate (about 10 litre/hour/tubule), the flow capacity of the irrigation network and yield from the source. An overhead storage tank (5m above ground level) was one simple solution to the problem, and one with which people were already familiar. A number of farmers had already erected tanks for other reasons, including the provision of water for home and livestock. But an alternative solution was to link the pump directly into the irrigation network, thereby eliminating the need for a tank. However, a pressure gauge was essential in this case (Figure 16). If the pressure was too high for the network then it could be reduced by diverting excess water into a cistern.

Filtration is essential as a protection against physical contaminants. In all five of the trickle systems installed, conventional, drip irrigation, plastic screen filters were used, with a capacity of 7 litres/sec. In two of the systems, two screens were mounted in parallel to minimise pressure loss. These filters, cleaned weekly to remove sand and organic matter originating from the well, proved sufficient to keep the systems in good order.

Manual gate valves were installed at the well head, either to divert excess pressure, as described above, or to close down the systems fed from a tank when the filters had to be cleaned. Other valves had to be used if it was necessary to block off one sector of the network and divert all the water to a second sector.

The components described above were common to all five systems installed by the Project for trial and demonstration purposes, two on the Project's farm and three on farms of interested local, small farmers (Hillman, 1984a). These basic components may be summarised as below:

- water source from a shallow hand-dug well;
- 5HP diesel pump, or comparable low pressure electric pump;
- pressure head from an overhead tank (5-6m) of 2,000 litre capacity, or equivalent flow direct from the pump with a gauge to monitor pressure;
- filtration using one or two in-line, screen drinking water filters;
- mains and sub-mains: 50mm PVC, commonly available in Oman, and buried to avoid ultra-violet light degradation;
- laterals and sub-laterals: 13mm alkathene imported from Dubai;
- emitters: 1.27mm microbore tubules, cut to length. Specially imported in 200m rolls;
- valves: manual gate valves locally available.

In addition to the above, trials were made with some more sophisticated components to illustrate how the basic system could be developed in the future. The three most interesting of these were a programmable timing control, 'fertigation' and a solar pump.

A programmable timing control box and solenoid valves were incorporated into the original trial irrigation network on the Project's farm. A simple shade protected the box and valves from direct sunlight but otherwise the control system, costing only a few rials, worked very satisfactorily without special care or maintenance for years. Once the system had been calibrated, exact quantities of water could be delivered through any sector of the network at pre-defined times. Thereby, no labour at all was involved in the daily task of irrigation, other than occasional spot checks to ensure that everything was running smoothly. However, it may still (in the mid-1990s) be too early to incorporate such timing devises into small-farmer irrigation networks. Until they have a better quantitative understanding of such things as volume, delivery rates and plant requirements they might leave it on a permanently incorrect setting.

Fertigation means the application of liquid fertiliser by feeding it into the irrigation water through the piped network. The Project developed a simple but effective method of fertigation and tested it on a local farm (the 'mixed farm', see below) with a mixed cropping of fruit trees and vegetables. A container holding a concentrated fertiliser mixture was plumbed into the irrigation main. The container was suspended sufficiently high above the main to exert a pressure slightly in excess of mains pressure. This ensured a slow and steady flow of fertiliser into the irrigation water, and subsequently to the root zone of the plants via the microtubules. The rate at which the fertiliser flowed varied with the height of the container above the main. As the container was translucent, the flow rate was easily measured.

The trial of the solar pump came about because of the Project's link with Shell International Petroleum Company (SIPC), and because it also had the local support of Shell Markets (Oman). A company in the Shell Group had integrated banks of solar voltaic cells with a slightly modified submersible electric pump to yield, from a known depth of well, 1-2 litres/sec water, sufficient to irrigate 1ha of mature fruit trees, or other crops. There were no other controls apart from the length of the day and the intensity of solar radiation, so the network naturally yielded more water on long, hot, summer days than on cool, short, winter days. The solar pump was linked to an irrigation network, of the type described above, installed in a one hectare square of land occupied by a mix of mature lime trees and other newly planted fruit trees. Extra water was directed to the mature trees by providing them with more emitters per tree. Until the end of 1985 the system was carefully monitored. Soil moisture levels were constant and appropriate, the mature limes trees gave good yields, and the newly planted trees grew successfully. Technical problems have arisen, but there was a more important financing problem: the solar unit and pump was much more expensive than the diesel pump and could only be installed and maintained by non-local expertise. So, even if the operating costs of the system were virtually free of charge it would take several years before it became relatively more economic than the diesel pump system. Thus a farmer would only be interested if he were able to secure a good financing package from OBAF. In practice, solar pumps will only have a role to play when such packages are readily obtainable and when the cost of the solar voltaic cells is greatly reduced.

As mentioned above, three of the trial and demonstration trickle systems were installed on local, small, private farms with the active collaboration of the farmers. One of the farms (here called the lime farm) was the set-up with the solar pump and the mix of mature limes and new trees described above. The second installation (the fruit farm) was for a square hectare of newly planted mixed fruit trees surrounded by a newly planted *Leucaena* wind break. In this case water was fed under gravity from an overhead tank into the network. The third installation (the mixed farm) irrigated an area of new fruit trees and an area of vegetables. In this case the diesel pump was linked directly into the network, and the headworks included, in addition to the filter, a pressure gauge and a fertigation unit.

It was an act of deliberate policy to undertake these early trials and demonstrations on privately owned farms, rather than on the Project's farm, once we had sorted out the basic principles. The farmers were well known to the Project, and it was made clear to them that although we hoped for good results the trials might not succeed. They accepted to work with us on these terms, knowing that at worst they stood to lose nothing but at best they would be left with the installation at no cost to themselves. Working with the farmers had the major benefit of their direct and active participation in the research and implementation processes (Okali *et al.*, 1994), including their ideas for modifying the design at the installation stage. We also believed that if they liked the results they would act as better demonstration and extension centres than we possibly could, through the normal course of conversation with their neighbours.

Unfortunately a doubtful start was made on the fruit farm. The farmer proved to be less interested than we had hoped, a situation made much worse by a highly localised and freak storm which threw down fist-sized hailstones which stripped the young trees of bark and leaves and cracked many of the partially exposed sub-mains. The farmer had neither the drive nor the technical capability to make the necessary repairs.

On the lime farm, however, there were two pleasing developments. First, the farmer's young son gave valuable assistance in monitoring soil moisture and other parameters. Second, the farmer, at his expense, expanded the network, this time linked to his diesel pump, onto his plantation of young date palms. In a sense this progress was too rapid. He knew that his limes were being irrigated very successfully but because that network used solar energy and worked throughout the day without his intervention he did not realise how much water was delivered. He had no intention of operating his diesel pump for the same length of time to irrigate his palms which were almost certainly not receiving sufficient water. But at least the farmer had demonstrated his belief in the system and that he was capable of extending it.

The third farmer, on the mixed fruit tree and vegetable farm, went one better. He quickly developed a good understanding of the different requirements of the system for vegetables and trees, he skilfully used the fertigation unit, he expanded the area under trickle irrigation, and he clearly saw the merits of the system for saving water as well as for saving diesel fuel and labour.

Unfortunately, the Project's irrigation specialist, Francis Hillman, left at the end of 1985 and could not be replaced until the spring of 1987 because of financial uncertainties associated with the change from PDO to MAF sponsorship in 1986. Important progress would have been made during 1986, including monitoring the three private farms, giving more training in the management of trickle networks, and providing one or more people with the equipment, training and general backing to prepare them for setting up small, local, trickle installation and maintenance enterprises. But at least, before his departure, Hillman had purchased stocks of piping and other equipment which the farmers could draw on. Thus, during 1986, without any support, the farmers maintained and developed their own trickle projects. But most interestingly, the owner of the mixed farm was contracted by the head of the local agricultural extension centre to install a trickle irrigation system for his field of lime trees. He fulfilled this contract without any additional technical training or assistance, and although his solution to the particular hydrological problems presented by the field was not ideal it was an enormously impressive performance considering that the area of technology was so new to him. It would have been easy to see him in the role of Oman's first small business in the field of 'Irrigation Services'. The principle had been demonstrated, albeit in a small way, that people from within the community could play a role in effecting the revolution in irrigation methods that Oman urgently needed if it was to check the rapid growth in water consumption by the agricultural sector. But such people would need recognition, they would need training, and they would need some help with the procurement of equipment not locally available. Then they would retain responsibility for the new irrigation systems within the community, making people interdependent and self-reliant. Moreover, the farmers would then know where to turn to if they had a complaint or if they wanted their networks modified or extended.

The work had also demonstrated, in a way convincing to the farmers, that the trickle irrigation system saved labour, time and money, for the following reasons:

- the land levelling and trenching needed for flood irrigation were not required;
- initial installation costs were no more than those for an equivalent network of irrigation channels;
- the only daily labour requirement was to start the pump and open and close valves at set times. In addition, a number of management practices (see below) were essential to ensure that the irrigation system was effectively utilised;
- weed growth was limited to those wetted areas near the crop plants; and
- fertiliser could be applied in the irrigation water.

However, these early trials also made it clear that the potential advantages could only be realised if certain management practices were carried out:

- filters had to be cleaned weekly;
- the network had to operate at the design pressure and duration in order to deliver the gross crop requirement of water;
- submains and laterals had to be flushed annually for several minutes;

- after an inter-crop break in cultivation it was advisable to operate the system for several hours in order to leach salt concentrations due to surface evaporation of moisture;
- it was advisable to make periodic checks on the concentration of salts in the soil;
- emitters had to be checked annually and either cleared or replaced if any blockage had occurred.

The installation costs of the system, as demonstrated by direct first-hand experience in the mid-1980s, were as follows:

- for 1ha of trees spaced at 8m by 8m and with six emitters per tree - RO503. This compares with an estimated RO688 using commercial trickle equipment plus the cost of a booster pump.
- for 1ha of vegetables in four blocks, row spacing 2m and emitter spacing 60cm - RO969. This compares with an estimate RO1,397 using a commercial system plus the cost of a booster pump.
- to the above costs should be added RO300-RO400 per hectare for rough labour, assuming that the farmer himself contributes to the work.

During the later-1980s and early 1990s the owner of the mixed fruit farm continued to expand and modify his trickle irrigation network. He was not happy with the fertigation unit with which he had been provided, so he designed and installed one which better suited his requirements. Having designed and installed it he was, of course, fully able to modify and maintain it. In 1994 MAF asked him if he would like to replace his system with the highly subsidised commercial system which they were promoting. However, he refused, even when approached directly by the Under-Secretary, saying that his system was better than theirs. In spite of this (or because of it!) MAF failed to recognise the man's potential. Relations between the farmer with the solar pump and the Ministry of Water Resources (MWR) have been somewhat more positive. They repaired his unit in 1994 and have monitored its activity since then.

MAF did however show some interest in the trickle system itself. In mid-1985 the then Directorate of Agricultural Research was seeking to establish forest nurseries at al-Kamil, in the Sharqiya in Northern Oman, and in Taqa in Dhofar. It signed a small sub-contract with Hillman, to advise on the 'design, purchase, installation, operation and maintenance of the al-Khabura type of microtube trickle irrigation and its application to forestry', to supervise its installation and to train staff in the design of future systems and their installation, operation and maintenance.

However, apart from the work undertaken by two of the three original Project trickle irrigation farmers there has been no active development of the low-head low-cost trickle programme since the end of 1985. By the time the Project's new irrigation specialist arrived in the spring of 1987 there were other priorities at both project and government level. On the Project's farm the systematisation of the flood irrigation system was required, indeed it was long overdue. Also, in response to the general increase in technological know-how in the country and, more directly, to the decision made by MAF to install sprinkler systems for Rhodes grass on the 300 farms of the GMDP (Ch. 5.3.1), the Project decided to convert part of its flood irrigated forage area to sprinklers.

4.9 A rural enterprise: fibre-reinforced-cement channels

Unfortunate early experience with earthen channels and polythene liners (Ch. 4.8.1) had at least suggested a set of design criteria for a new type of liner. The design criteria included: strength, resistance to decay, manufacture from local (or at least locally available) raw materials, capability of local manufacture and repair in a 2-3 man workshop, mobility and space-saving. The capability for local manufacture was regarded as very important because it would help fulfil the important project community aims of increased local responsibility and mutual self-reliance based on interdependence between different local economic groups, in this case workshop technicians and farmers.

In 1979 the Project turned to the Intermediate Technology Development Group (ITDG) in the UK for advice and assistance in turning the design criteria into the product we wanted. For local materials they recommended a dryish mix of cement and sand reinforced with natural fibre and the manufacture from this mix of 2m-long channel liners (here called sections) with a wall thickness of approximately 1cm. These channel liners could be made and cured in a small and cheaply equipped workshop. It was hoped that all the other design criteria would also be met, on the basis of previous ITDG experience with fibre-reinforced cement (FRC) adapted to suit local conditions and requirements. The fibre had a fourfold value in the cement: to improve the tensile or flexural strength; to improve the impact strength; to control cracking; and to change the flow characteristics of the sand-cement material in the fresh state (Winstanley, 1983; Hillman, 1986a-b).

The cement, purchased locally, was imported from a cement works in the UAE. The simplest and cheapest sources of sand were the beach and local wadi beds. It was not ideal, being composed of smooth round grains with too high a proportion of fines and some shells, stones and other debris, so it had to be sieved. For fibre we turned to the date palm because at the base of each frond is produced a sheet of matted fibrous material (traditionally used for making rope, as we had seen). The sheets were passed twice through a specially imported cutter (a Hunt chaff cutter from the UK) to yield fibres of 20mm-30mm length, and received no other treatment. A Benford Half-bag cement mixer (locally purchased) fitted with an electric motor was used to mix the ingredients. This was achieved without balling up the fibre, and greatly increased the speed and reduced the manual effort of hand mixing. About 600gm of fibre were required per section.

The principle of section manufacture was to plaster a thin layer of FRC over a rectangular polypropylene sheet, suspend the sheet from its long edges so that the plaster hardened with the cross-section of a narrow catenary curve and cure it while still saturated with water.

Shaping a section required three simple items of equipment: a suspending frame, a laying-up board, and stands. The frame consisted of the rectangular polypropylene sheet, mentioned above, clamped between laths fastened to its long sides, and a second sheet of polypropylene clamped between the same laths and folded over the FRC as a top sheet. The laying-up board was made of three layers of 10mm plywood 2.1m long. The successive plywood layers were each slightly narrower, so forming two steps along either side. The first step located and held the suspending frame in place. The second step produced a thickened top edge, for greater strength, on either side of the channel. Wooden battens at the head and foot of the laying-up board located the end-pieces which gave the channel sections neat straight ends. The stands for the suspending frame were set in concrete far enough apart (2.25m) to allow the laying-up board to fit between them but close enough together to allow the laths (2.4m) of the suspending frame to be supported by them when they were lifted, with minimal disturbance, from the laying-up board.

Tamping was a very important part of the process of levelling the plaster in the suspending frame. It was essential to remove all the air bubbles which otherwise would have weakened the section. After tamping, the top sheet was carefully folded over the plaster. It prevented slumping when the frame was lifted onto the stand, and it helped keep the plaster saturated during the 24 hours that the section hardened on the stand. After one complete day the sections were then demoulded, when they were strong enough to handle, and put directly into a water tank for two days wet curing. From the water tank they were moved to a shaded area where they were covered by hessian cloth kept constantly damp for a minimum of ten days. The channel sections were then put outside for storage until needed.

The first sections manufactured by the above process (the so-called Mark 1 sections) were 2m in length, 10mm thick and slightly tapering. They were also deep (over 300mm) in comparison with their width (about 220mm), which made them more tolerant of unevenly sloping ground and made it easier to start and maintain siphons. A 1:2 sand:cement ratio was used with 1.25% by weight of date palm fibre. The total weight of materials came to 20kg.

Most of the work was done by two Omani workmen on the Project's staff, but in practice the production control was insufficient so the channel sections varied in weight from 20-35kg. Production was halted when hairline cracks began appearing in the installed channel sections. In theory the taper on the sections allowed the narrow end of one to slot neatly into the wider end of the next. However, lack of uniformity in the section profile made it difficult to obtain a good joint. Failure usually occurred where water, leaking from a joint or crack, washed away the earth around the channel leaving it unsupported.

The cracking and jointing problems of the Mark 1 sections were the subject of study by ITDG in the summer of 1981. In the autumn of that year recommendations to overcome them were implemented when the new production run started. The tapered cross section was replaced by a constant cross section so that the sections would butt joint together, the top 50mm of each side was thickened to 20mm, and both the suspending frame and the stand were made more rigid to eliminate any possibility of sagging. It was also decided to experiment with a 1:1 sand:cement ratio as well as the 1:2 ratio, hoping that having less cement would reduce shrinkage and the tendency for cracking.

Two methods of installing the butt-jointed sections were tried: into the ground, or above ground supported on concrete stands. In both cases levelling was achieved with the help of a taught string. The joints between the sections were sealed using a bitumen emulsion[15] mixed with date palm fibre, but it was easier to make a permanent waterproof seal when two sections were butt-jointed across a concrete stand. The stands also reduced the work required to level a channel down an irregular slope.

Discussions with farmers had revealed a lack of preparedness to familiarise themselves with the use of siphons. They were used to channels with gates, albeit mud channels with mud gates, and were reluctant to trust a siphon or to acquire the knack of starting one. This posed a question of fine developmental interest. Should we accept the perceptions of the farmers that the siphons were not suited to their needs, or should we persevere and persuade them that the combination of siphons and channels without gates had certain real advantages of which they were not aware simply because they had no experience of them? They advantages we saw included retaining the simplicity of the product, flexibility of water off-take point, and therefore more even spreading of water, reduced erosion, no leakage from the channel and no need to open and close gates. These advantages amounted to a more efficient use of time and water but in the early 1980s, time on small-farmer holdings was not at a premium and there was no awareness that water was becoming a scarce commodity, therefore the advantages, if real, pertained more to the future and were unlikely to gain quick recognition. The advantage of simplicity would be reflected in a reduced product price but as we were still operating at a trials and demonstration phase, sales had not come into the question.

In practice we resolved the dilemma by the obvious compromise. We made some sections without gates and some with. On the Project's own farm we continued with the use of siphons.

The Mark 2 sections which had a lean cement mix (1:1 sand:cement ratio) and were left longer under damp cure and in the shade seemed to remain almost free of cracks before installation. But after installation, and after exposure to an

---

[15] Flintkote No. 5 obtained from Shell Markets (Oman).

Omani summer, cracks had appeared in most if not all sections though where the sections were lying firmly on or in the earth little leakage occurred and the channels continued to function.

Mark 2 sections installed as channels on Omani farms suffered less cracking. They were installed literally as liners, earthed up to the top, which will have given them more support and kept temperatures more uniform. All these channels were fitted with gates.

In the spring and summer of 1983 the Mark 2 catenary shaped sections were under continual production at the rate of four per six-hour day (Hillman, 1983). But although the problem of cracking was not serious enough to halt production or cause unacceptable leakage (where the sections were well earthed up) it had serious implications for the public acceptability and therefore commercial viability of the product. Moreover, the lack of enthusiasm for siphons had still not been overcome. In consequence, a series of interrelated design and production decisions were taken to address these questions, and to produce a Mark 3 section. New production procedures intended to strengthen the sections and eliminate cracking were suggested as a result of the destructive testing of sections at Durham University. They included: washing-winnowing the sand to remove silt and fines; vibrating the mix in the flat stage; and positive location of the suspending-frame laths in the stands to avoid uneven hanging and stresses.

A technique was devised for washing the sand but although this indeed removed large quantities of dust, subsequent comparisons showed that it did not significantly strengthen the sections and so washing was discontinued. A pneumatic vibrator attached to the tamping bar did, to some extent, facilitate spreading the plaster but it did not have a noticeable effect on the strength of the newly designed Mark 3 section, and its use also was discontinued. But the more positive location of the top bar facilitated the production of a standard section, and was continued. In another change, the wet cure tank was enlarged so that sections could remain submerged in water for a 48 hour curing period. The damp cure shade area was also extended. Section strength was ensured by the simple expedient of thickening the section wall to a uniform 20mm, and the tendency to shrinkage and cracking was reduced by moving to a 1.75:1 sand:cement ratio. The total materials for one section (and one gate collar) were 40kg sand, 28kg cement and 3kg date palm fibre. Water content was controlled by experience, but approximately 10 litres/section was required; a compromise had to be met between maximum strength (dry) and practical workability (wet).

As for design, the original high, thin catenary shape, suited for use with siphons, was replaced with a semi-circular shape when it was decided exclusively to use gates instead of siphons. The disadvantage of the move to gates, however, was the loss of the original concept of a simple one-element product wherein the workshop had only to manufacture sets of identical 2m sections. The Mark 3 design required sections, joining-sockets to seal and support butt-jointed section ends, stands and gates. The gate design included the sealing of a polythene elbow joint, further complicated with a removable bung at one end and a length of discharge pipe at the other, into the FRC gate collar.

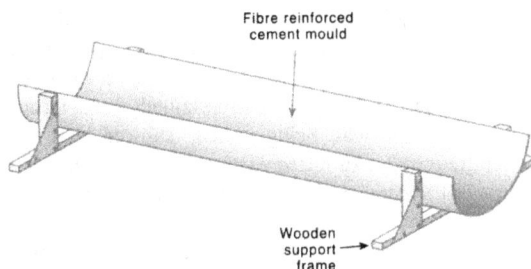

Source: Hillman, 1986a

Figure 17. Mould for Mark-4 channel liner section

The general method of manufacture for the Mark 3 semi-circular section was the same as before except that the polypropylene sheet onto which the FRC plaster was spread and tamped down, was carefully lowered into a mould made from a PVC sewer drain, available in Muscat, cut in half and supported in a simple wooden cradle (Figure 17). As before, the inner face of the plaster was covered with a second plastic sheet to slow drying.

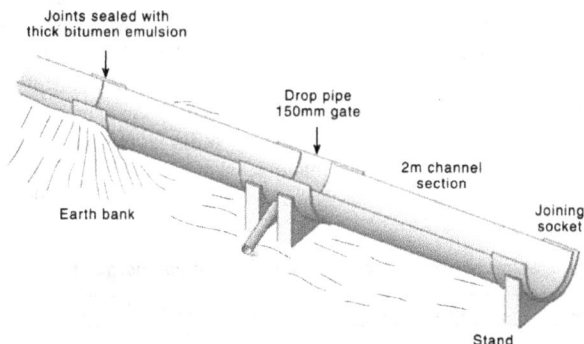

Source: Hillman, 1986a

Figure 18. Mark-4 channel liner assembly

The joining-sockets, using a similar process, were moulded over the outer edge of a section.

The gates were similarly moulded over a section but both the wet FRC plaster of the gate and the mould had holes cut in them. A PVC elbow was placed through the holes and cemented into the gate (Figure 18).

137

The stands were made of ordinary concrete poured into a box mould designed to make six at a time.

All of these items which together made up the prefabricated channel assembly were left overnight to set, and carefully demoulded the next day. After being left in the wet cure tank for three days they spent at least one week in the damp cure shaded area where they were hosed down each day.

Subsequent experience and suggestions obtained from farmers in whose fields the Mark 3 design was being tested, showed that some Mark 4 modifications were required. An intermediate size between the original 300mm and 400mm diameter PVC sewer pipe moulds was wanted, and it had to be rigid to prevent warping. Both of these requirements were met by the use of a 350mm steel pipe cut in half lengthwise to act as a template over which FRC semicircular moulds were cast. In that way any number of virtually identical rigid moulds could be made cheaply.

Installation methods were altered to match the changes to the channel assembly, and were refined through practice. Experience showed that a slope of channel of 0.25-0.5%, which generally conformed with the slope of the land towards the sea, best fitted the combination of channel hydraulic characteristics and the local 6-15 litres/second range of water delivery rates from the diesel pump-sets. At the head and tail of the channel, stands with joining-sockets on them were positioned with the help of a surveyor's level, and intermediate points were similarly fixed if the channel was very long. Builder's twine was stretched very taught between them, and additional stands and joining-sockets positioned every 20m and then every 2m along the twine. Gates were positioned where required. Beads of bitumen were daubed onto the joining-sockets, and the sections laid in place. Installation, as can be seen from this description, was not a simple process that could be learned and followed by the average small farmer. However, it was envisaged that the necessary skills would be learned by local channel laying teams, just as the manufacturing skills had been learned by workshop employees, and that the teams would be employed by the farmers to undertake the work.

Table 11. Installation of irrigation channels on small-holdings, 1983-7

Item	1983	1984	1985	1986	1987[1]	Total
No of channel lengths	495	905	1917	1616	1076	6009
Gates	n/a	n/a	n/a	584	169	
Stands	n/a	n/a	n/a	n/a	661	
Sockets	n/a	n/a	n/a	n/a	749	
km of channel	1.0	1.8	3.8	3.2	2.2	12

Source: Dutton, 1986a; Ann Reps, 1986 and 1987

(1) 8 months only
n/a = not available

In practice, by the autumn of 1984 all the manufacturing and installation operations were in the hands of Omani technicians from al-Khabura who worked with growing skill, confidence and interest. Many hundreds of channel sections had been installed, including 640 (1,280m) semi-circular sections, and there were constant orders for more (Table 11). Thus, with the benefit of continual discussion with the technicians and the farmers, the Project had arrived at an apparently acceptable pre-fabricated irrigation channel manufacturing and installation process. However, the workshop was on the Project's farm and all the technicians were project employees.

How were we to get the channel project truly into the local community so that it could continue even if the Khabura Project ended? One starting point was that since the early 1970s two to three-man workshops making concrete blocks had appeared in almost all the coastal villages. These workshops, established to fulfil the new and growing demand for concrete block houses, were the result of local, small-scale, private entrepreneurial initiative, without any government assistance. The owner, a local man, purchased a hand-operated press (often imported from India), built a simple *barusti*-roofed shade area, purchased cement, found a source of sand, used his own water, and needed sufficient open space to store the blocks before sale. Typically his labour force was Indian or Pakistani. The success of these workshops indicated that sufficient organisational skill and entrepreneurial flair, and experience with cement-based products, existed locally to form the basis of the kind of workshop we hoped to create.

One of the workshops was owned by a friend of the Project, Khamis bin Ibrahim al-Balushi, and he seemed to have the appropriate combination of track record (successful farmer, block maker and house builder) flair and determination to make a success of the channel enterprise. By this time he was thoroughly familiar with the channels, and he expressed his commitment to the new channel workshop project. With his backing we prepared a small business plan for a workshop with a production level of 12 units (sections, sockets, and stands) per day. Altogether the building, plumbing, equipment, initial stock of materials and labour for building came to RO1,500. This would yield a net income of RO6 per working day for the owner on the following assumption: 290 days work per year at full production level, loan repayment over three years and a sale price per unit of RO2.2. If the channels were subsequently installed at RO1 per section, the total installed cost would be RO3.2 per 2m section, or RO1.6 per metre.

On the basis of this business plan Khamis Ibrahim secured a loan of RO1,500 from OBAF, and the workshop was in production by the winter of 1984/5 (Hillman, 1985). In practice the three expatriate Indian workers produced 10 units per day (not the planned 12) but each unit fetched RO2.25 and Khamis took a profit of RO4.5 daily. But Khamis was not really prepared to put a great deal of effort into making a long-term success of the venture. He probably, for example, could have achieved a higher rate of production at a lower labour cost if he had been more actively committed. He continued to rely too heavily on the Project's staff for securing new orders for channels, and he insisted that the

Project would have to buy the sections from him for onward sale to the farmers or else the farmers would delay payment until the workshop was forced out of business. In fact, this was a real problem and would have to be faced by any other type of small enterprise struggling to establish itself locally. Nevertheless, the channels were in demand and Khamis was making a profit even if the enterprise, like so many others in Oman, depended on cheap, imported Indian sub-continent labour and was not therefore giving employment to or teaching new skills to Omani trainee technicians.

But although the workshop was profitable the channel installation operation did not cover its costs. The work remained in the hands of the Omanis trained by the Project, who by then were doing a thoroughly professional job but at a much higher wage than would have been earned by a technician from Pakistan. The actual installation costs were RO1.5, but only half of this was charged to the farmer. The Project carried the difference in the hope that as the channels became more widely known and in greater demand the charge to the farmer could rise or other workshops would open in competition and operate more cheaply. Certainly demand for the channels was rising. The number of sections installed rose annually from 495 to 905 to 1917 in the three years up to the end of 1985 (Table 11), by which time they had been purchased by 69 farmers in 22 villages, and remained at that level until the workshop closed.

As the farmers, with their new diesel pumps, continued to expand their crop areas and therefore the length of traditional earthen channels of their farms they became very aware that large volumes of water were being lost by them - though it is important to note that when the farmers were asked why they liked the channels, the reply was that they saved time, diesel fuel and wear and tear on the pump - not that they saved water!

The channels were also being tested in other parts of the country. As early as March 1983 Nizwa Agricultural College purchased a line of the Khabura Project's fibre-reinforced channels and installed them on a part of its farm which it had dedicated to the demonstration and comparison of alternative types of irrigation systems, from the traditional to the high-tech. Also, in mid-1985 the then Directorate of Agricultural Research was seeking to establish forest nurseries at al-Kamil, in the Sharqiya in Northern Oman, and in Taqa in Dhofar. For the site at al-Kamil the Ministry bought 200 lengths of the Khabura Project's fibre-reinforced channels.

However, another alternative to earth channels had been devised. Some farmers, using cheap Indian labour, were laying channels which had a cement floor, and walls of concrete blocks. The immediate advantages of this system were that the blocks were cheap and had become, as mentioned above, ubiquitous, and they could be cemented together in long parallel lines without much skill to form the channels. Gates were sometimes as simple as a gap between two blocks, that could be closed with a wedge of rags, but more typically cement grooves were shaped to act as guides for a rectangle of wood.

The longer-term disadvantages of the system were that the channel once laid was immovable, the crudely made concrete blocks tended to erode under the pressure of regular wetting and drying, and the gates invariably leaked, sometimes copiously. But short-term advantages tend to outweigh longer-term benefits when people are deciding on a course of action, and the cheap and cheerful concrete block channel prevented an expansion of the FRC channel production system to the point where the whole enterprise would have become truly self-supporting.

Khamis Ibrahim's workshop continued in operation throughout 1986 and much of 1987. However, it still had not become truly independent of the Project and Khamis was losing interest because his other business ventures were failing, particularly house building where he seemed unable to secure payment of large sums of money owing to him. He managed to pay off his loan from OBAF but, finally, he had to sell his farm and was therefore forced to stop channel production. The FRC channel project had come to a close. But why had the project failed after it had moved so far down the road towards success? A number of reasons may be cited:

- Workshop owner: Events showed that we had made a poor choice of workshop owner. Although he seemed to have the right combination of active interest in the Khabura Project, experience with a concrete block workshop, other successful enterprises including farming, finance, and position within the local society, this image was more apparent than real. We discovered, too late, that he was too dependent, not sufficiently committed or energetic, was worried by other weak and failing businesses, and did not carry sufficient authority either to carry conviction about his product or to collect his debts. But we partly sought his participation because he occupied a middle-ground position locally. We were reluctant to support someone who was overtly wealthy and successful (it did not seem to be what the Project was about) though we realised that no progress would be made by trying to work through one of the humbler members of the community. We probably should have tried to interest a son of the senior local shaykh, a man who held a high position in central government, had obvious influence nationally and locally, and owned a thriving local farm on which he installed a large number of the Project's channels.
- Small business environment: Prior to the start of oil exports in 1967 the central government had no money and offered no support to local enterprises. Yet many small craft-industrial and service industries existed and, within the general national economic constraints, were successful. Oil wealth gave a tremendous boost to the public sector and although, in more recent years, systems of support and encouragement have been created for the nascent, formal industrial sector, longer-established rural enterprises, with the exception of fishing and aspects of farming, have tended to whither away. Their potential role as growth points for a renewal of rural enterprise had not

been sufficiently appreciated, and so the small business environment needed for their success did not (and does not) exist (Kinsey, 1987b).

- Living on credit: A complex pattern of rural indebtedness, or living on credit, has been a feature of village life. People, therefore, do not have scruples about the prompt payment of bills and it is socially difficult to put pressure on them to do so. This was a real and major problem when trying to make the channel project financially independent; the workshop owner was convinced that without our role as intermediate channel buyer many farmers would not pay and his business would fail for reasons of liquidity.

- Expatriate labour: By the 1980s practically all labouring and technical jobs in rural Oman were taken by Indian or Pakistani immigrant labourers, sometimes, as in the new garages, because they had the expertise, but always also because they were much cheaper to employ. This placed the staff of the Khabura Project in a dilemma because they were in Oman to develop rural skills amongst the Omani population. Yet the channel enterprise was unlikely to compete effectively with the alternative block-built channels if the former used expensive Omani labour and the latter cheap Indian labour. In practice we compromised; we did not object when the workshop owner used Indians to make the channel sections, but we employed Omanis to install them. In doing so, we raised the installers to a high level of technical competence but left them without a job when the enterprise failed. Perhaps we should have concentrated single-mindedly on ensuring the success of the enterprise, hoping that in the longer-term Omanis would replace Indians as national pressure for Omanisation grew.

- Siphons or gates: Although the design of the channel system had undergone a steady evolution in response to local experience and small-farmer ideas, room for improvement still existed. But the project decision to opt for gates instead of siphons as the sole means of delivering the water from the channel to the field, a decision made in response to locally expressed preferences and in accordance with accepted developmental philosophy, may have been wrong. It made the channel assembly considerably more costly to make and install, but cost was not taken into consideration when local farmers were expressing their preference for gates, which was simply the system with which they were familiar. In retrospect, more conviction should have been retained in the longer-term merits of siphons, and much more effort should have been expended to develop a siphon (specially shaped, perhaps) which would have functioned easily within the semi-circular Mark 4 channels.

- Flood or overhead irrigation systems: When in 1976/7 the decision was made to opt for an improved system of flood irrigation rather than adopt an overhead irrigation system it was probably justified in terms of local experience and available support. However, this decision did not result in an operational channel workshop until 1985. And by the time the workshop closed, Oman as a whole had gained much more experience of piped irrigation systems of all types, and the government, now much more aware of the urgent need for water conservation, was losing interest in any form of flood irrigation. This does not mean, however, that there was no place for flood irrigation. Subsequent evaluation of sprinkler systems, whose installation was being promoted and subsidised by MAF, showed that typically they were being used very inefficiently (Stephens, 1993a). Flood systems can be an efficient way of irrigating field crops on small farms.

In conclusion, in spite of the changing realities in rural Oman, and in spite of the fact that the original channel project failed to thrive, irrigation channels still have a role to play. A new project could be built on the experience gained from the old one if proper attention is given to the problems identified above. Although high pressure piped irrigation systems are now being encouraged their introduction is costly and should proceed with caution. Some channels will be in use in many Omani small-farms for years to come, and they should be made as efficient as possible.

From the channel project can be drawn lessons of broader relevance to rural development in Oman. The problems of identifying potentially effective entrepreneurs, of creating an appropriate business environment, of expatriate labour and of payment for services rendered, will be common to all new, rural, small-scale ventures in the country, and will need to be resolved.

4.10 Sheep and goat pens

We involved the Project in the design of various types of farm building, the most important of which were the pens for the sheep and goats.

4.10.1 Design criteria and improvements

From the outset, the design criteria for the livestock pens were kept simple: provide shade, restrict freedom of movement to a defined space, allow free passage of air, permit easy feeding and watering, allow easy removal of waste products, provide a dry area in case of rain, and facilitate easy livestock handling. We restricted ourselves to using materials of local origin, or at least available on the local market. Construction relied on tools (hammer and nails), easily within the financial reach of small farmers, and skills that tradesmen or the farmers themselves could readily master. More sophisticated housing systems (ICAE, 1987) would have been inappropriate for reasons of sustainability.

The pens were built in lines concentrated in a stockyard. This minimised the space required (leaving more land for the forage crops), saved on material and construction costs, facilitated access to the animals and allowed neighbouring pens to benefit from each others shade. But it increased risk from ticks, flies, fleas and other parasites and disease micro-organisms. The pen size, for an improved line of six pens built in 1984, measured 5.25m by 7.75m (Figure 19), with the feed racks along the short side (Winstanley and Hillman, 1984). New pens designed (but not built) in 1989 (Figure 20) were larger but equipped with removable hurdles so that groups of animals could, with minimal distress, be confined to small areas for ease of handling.

The pen frames were made of 2 inch by 4 inch imported timbers and chain link fencing (both available in al-Khabura *suq*). The roof timbers were covered by traditional date palm frond *barusti* sheets and the animals stretched their heads through the front of the pens to reach forage placed in a manger. The design was efficient in that it was not necessary to enter the pens to supply feed. However, with usage, design faults manifested themselves leading to a succession of improvements to water supply, flooring, fencing, feeding, protection against the elements and lighting, as discussed below.

Initially water was supplied in buckets filled by hose pipe and left in the pens. But the virtues of this cheap simplicity were outweighed by the disadvantages of buckets being easily knocked over or being fouled by dung and straw. In summer, when the animals needed constant access to water, replenishing the supply was labour intensive with the constant danger of leaving the livestock under stress through water shortage. A major improvement was made possible when, as a result of oil wealth starting to flow into al-Khabura village, new houses had plumbed water supplies for the first time, and Indian and Pakistani plumbers settled into the village to undertake the work. Plumbing expertise, piping and fitments therefore became available to the Project's staff which designed a livestock watering system based on a tank with a float valve so that cool water constantly flowed to a series of small, shallow, concrete water troughs (Figure 19) which never dried and were easy to clean (Winstanley and Hillman, 1984). The system rapidly paid for itself, as a result of labour saving, and developed another set of skills within the community, albeit, at that time, in the hands of South Asian artesans. The system stood the test of time (Barker, 1991) though with the later improvement of a locally made fibre glass header tank manufactured by one of the companies that fabricated hulls for the new government sponsored fishing boats. This option was cheaper, supported another new local industry which was seeking to diversify, and kept the water cooler.

Originally the floors were of earth, for cheapness. However, the process of cleaning the pens inevitably removed some of the earth which lowered the floor and left it increasingly uneven. Rainwater thus flowed into the pens and created muddy pools which exacerbated the ever-present pneumonia problems and generally left the animals unhappy and stressed. We therefore cemented the part of each pen which was under the *barusti* roofing, and experimented with straw and wooden sleeping platforms to create a more pleasant sleeping environment. It was much easier to remove the dung and straw from the cement flooring, and to keep it clean and dry in wet weather.

The chain link fencing was of poor quality and low tensile strength. It was sufficient to contain the docile local sheep but the goats tended to walk and lean against it, or butt it, or work their horns behind the wire and pull, and thus unravel it. A horizontal bar at adult shoulder height greatly reduced the problem, as did alternating goat with sheep pens; the goats were much more destructive if they faced other goats directly through the fence. The fencing and roofing timbers were unseasoned and susceptible to warping, rotting and wood boring pests. Wood preservatives were not locally available and the only defence against woodrot was to set the uprights into cement, which also strengthened the whole pen structure.

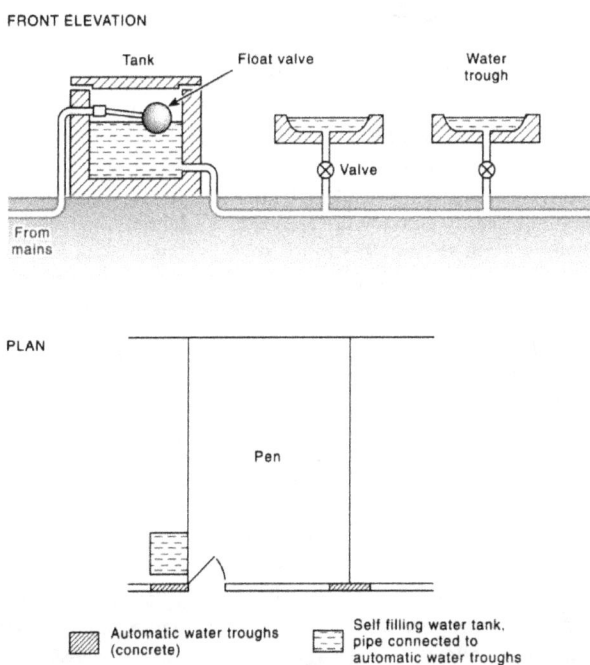

Source: *Winstanley and Hillman, 1984*

Figure 19. Automated livestock watering system

For feeding, the solutions resorted to were always a compromise. If hay and concentrates were provided in racks and bowls in the pen, the animals had plenty of space so that competitive stress was avoided when eating but the system was labour intensive and led to fouling of the feed in the bowls. If the feeds were provided from outside the pens, their supply was quicker and did not disturb the animals. However, both goats and sheep were more stressed by having to compete for eating space and it was difficult to prevent young animals, particularly goat kids, from escaping. A pilot structure which included a wire mesh food rack and horizontal sliding rails was effective and also developed the metal-working skills of a local workshop. However, it was possibly too elaborate to be emulated by local farmers, even on intensive goat and sheep holdings.

Protection from the sun and rain was essential. The *barusti* roofing gave excellent shade from the sun. If other considerations permit, it is ideal to construct a line of pens with an east-west orientation so that only the end pens are exposed, respectively, to early morning and late afternoon sunshine. For similar reasons the roof should decline to the south. *Barusti* blinds were sometimes hung from the roof to provide extra shade, but the goats were tempted to gnaw and destroy them if in reach. Pen design had to allow for rainfall because although it was an infrequent occurrence the smallest amount led to problems. The hard packed soil in the pens inhibited drainage and the pens became puddled and muddied. Therefore, where possible, part of the floor was concreted, or sloped to remove the rainwater. We also found it advisable to use a double sheet of *barusti* sandwiching a polythene layer for roofing, to prevent the roof dripping onto the unhappy animals beneath it. Many pens were provided with wooden sleeping platforms so the animals could lie down above ground level, particularly in wet weather, and thus gain some protection against pneumonia. We also found it very beneficial, as a protection against the effects of wet weather, to cover the main farm paths with a layer of wadi gravel, raked and rolled to a depth of 2-5cm. Movement of animals, people, barrows and forage carts was greatly facilitated.

Good lighting was also found to be essential because the animals had to be checked at night, particularly during the breeding season. We used sealed units for safety but in practice local farmers started to use cheaper open bayonet light fittings.

A single 9m by 6m pen to hold 10-15 adult sheep or goats plus young, based on the design criteria discussed above, cost RO221 in 1983 (Winstanley and Hillman, 1984). The line of six pens built in 1984 allowed substantial savings per pen, and in fact cost only RO1,100 though they housed up to 90 adult sheep and goats and their followers. A design for another line of pens was discussed in full detail with a small building contractor in al-Khabura in 1989 (Figure 20). He provided a full breakdown of costs, and signed a contract with the project on this basis for a total of RO1,687 (Barker, 1991). We were very familiar with his work and had every confidence that he would have done a good job but because the Project closed in that year the contract had to be cancelled. For comparison we had also obtained quotes from recognised companies for a range of buildings they were prepared to erect to do the same job. The least, for just the frame and roofing, was RO4,000-5,000 if the frame was of wood, and over RO10,000 for a steel portal. Thus our preferred solution would have given us a building complete in every detail at the fraction of the cost of using a recognised company to provide only the frame and roofing. At the same time it would have enhanced local construction skills in a small company that was readily accessible to all the local livestock farmers. When MAF subsequently established the GMDP they employed recognised companies to make the pens at great expense and with minimal (or no) benefit to local enterprises.

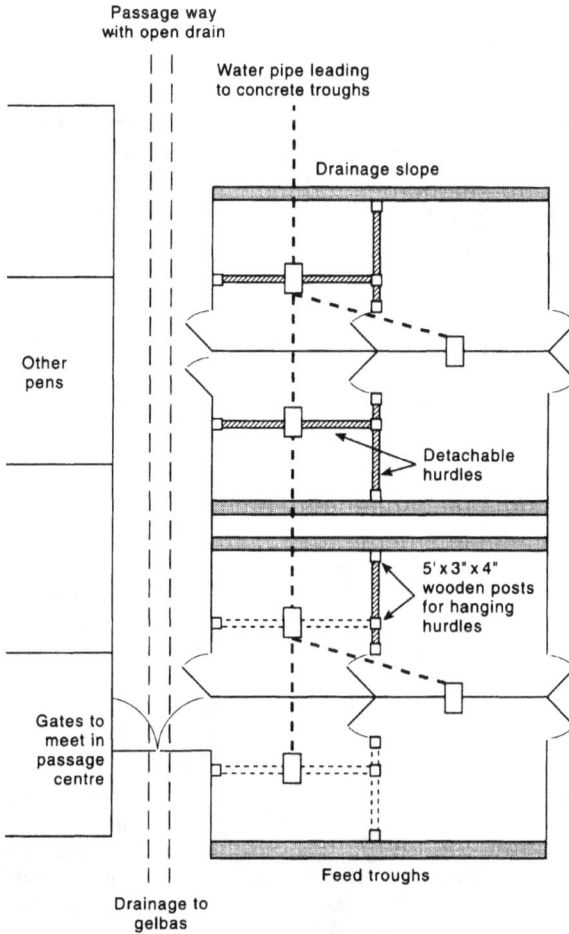

Source: Barker, 1991

Figure 20. Plan of proposed livestock pens, 1989

4.10.2 Contract pen building

The Project's pen design criteria emphasised simplicity and the use of locally available materials. Nevertheless, the size of our enterprise and the fact that trialing and demonstrating improvements was part of our brief, meant that the local farmers tended to view the Project's standard pens as too large, complex and costly for their needs. However, by the early and mid-1980s many farmers were taking onto their own farms improved local and cross-bred sheep and goats which needed pens. They could see the management merits of improved housing for these animals, so, in discussion together, we evolved pen designs suited to smaller numbers of animals owned by families where labour saving was not a priority (Barker, 1991). The pens were simple and cheap, and could be constructed by local labour. While the designs were evolving, the Project provided the pens free of charge. Under this system some construction began in 1982/3 but most units, in practice provided for our so-called participant farmers, were made during the years 1986 and 1987, as shown in Table 12.

Table 12. Livestock pens constructed for participant farmers, 1986-7

Year		Activity
1986:	Jan-Mar	8 new units constructed, plus running repairs on others
	Apr-Jun	4 new units constructed, plus running repairs on others
	Jul-Oct	1 new unit constructed, 3 enlarged, plus running repairs on others
	Nov-Dec	1 new unit constructed, 3 extended, plus running repairs
1987	Jan-Dec	11 new units constructed, 4 extended, plus running repairs

Source: Ann Reps. 1986 and 1987

Later, partly from principle and partly because of cash shortage we attempted to devise a system whereby the farmers paid for the pens, and the construction work was done by Omani tradesmen, trained by the Project. An employee of the Project, who was also a local farmer, grasped this opportunity and was prepared to use his own pick-up truck to buy the materials from local suppliers. He was also skilled in pen building because that had recently been his main employment on the Project's farm. He was also well versed in the merits of using pens and was thus a good advocate for them.

The two initial contracts were negotiated by the Project on behalf of the employee-farmer, in his new role as 'contractor'. Both were for single pens to hold 6-10 adults. Both made partial use of some structures and materials available on site so it was not possible to calculate costs according to a simple formula, based on area for example. The cost to the farmers was RO20 and RO55 respectively, the latter including hay racks and other extras. Subsequently, the contractor prepared quotes based on the estimated time required for each pen contract. Materials were supplied by the 'client' or, for a small charge on top of cost, by the contractor.

Because the pens were constructed during normal project working hours the Project charged the farmers for the time that the contractor was away from his normal duties. The minimal charge was RO1 per hour though wherever feasible this was raised to RO2 per hour which represented a profit to the Project.

Unfortunately the cost and system of payment initially killed interest from the poorer farmers primarily because they had to pay in a lump sum. All farmers, in accordance with local custom, expected a period of credit, and it was difficult for the contractor to press them for payment. The problem was only alleviated when the contractor was able to shame farmers into paying by saying that the Project was demanding repayment for the labour charge. In practice we were then able to insist that the farmer paid the labour charges before work began. In some cases the Project also acted as arbitrator where there was a dispute.

Under this system there were a lot of developmental gains: many pens were built according to joint design inputs by the farmer and the contractor, construction skills were enhanced within the village community, workshops were given sub-contracts to supply materials and manufactured goods such as hay racks, and the sheep and goats were better housed and managed. But the flaw lay in the casual local attitude towards debt repayment and the seeming lack of any local system of enforcement. We could not find a way to hand responsibility for organising this debt repayment back from the Project into the community.

4.11 Locally made tools and equipment

An intensive small forage and livestock farm such as ours offered endless opportunities for the innovative design of tools and equipment that could be manufactured in the small workshops that were newly opening in al-Khabura and in the other larger settlements in rural Oman. Primarily the workshops opened to repair the ever growing number of cars. They were able to diversify as the local demand for 'modern' housing with electricity, piped water, furniture and drainage systems grew. We were able to make use of this growing technical capability - albeit as with the plumbers we had already used, mainly in the hands of tradesmen from India and Pakistan - for the manufacture of tools and equipment for field and livestock work. Some ideas originated from the Project staff, others from local farmers. All were tried, tested, modified and improved over a period of years. Almost all the equipment required basic metal working and welding skills in its manufacture but as the majority of small towns in Oman have a welding shop the tools could be made locally at minimal cost. Hand tools for handling the forage crops included a hay fork and a rake. Equipment to transport forage included a basic wheel barrow with a wire mesh extension and a twin-wheeled transporter (wheels and axle from a scrap yard). Concentrate feeders were made from wooden planking or old plastic water containers. Stock hurdles and creep hurdles were welded from piping (Figure 21). Forage racks had a welded piping frame covered with 4cm weld mesh, and a youngstock weighing strap was fashioned from old meal bags, two wooden rods and some

CHANGING RURAL SYSTEMS IN OMAN

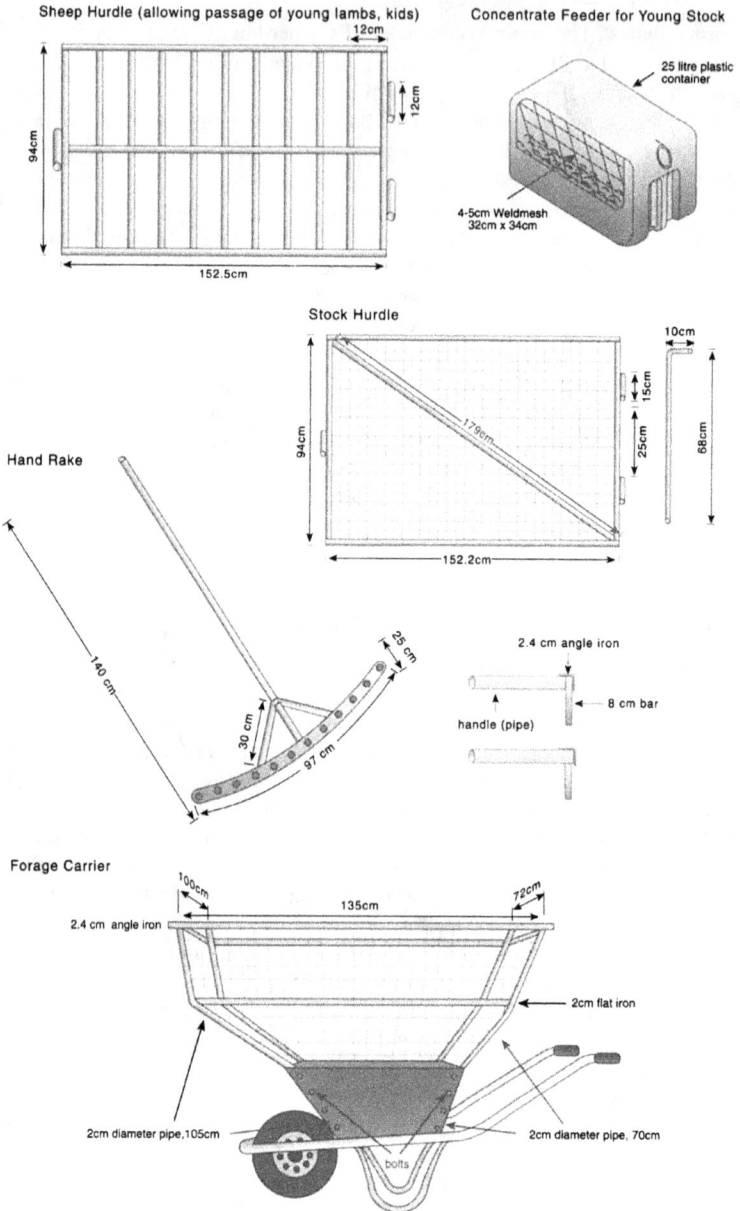

Sheep Hurdle (allowing passage of young lambs, kids)

Concentrate Feeder for Young Stock

25 litre plastic container

4-5cm Weldmesh 32cm x 34cm

Stock Hurdle

Hand Rake

2.4 cm angle iron

8 cm bar

handle (pipe)

Forage Carrier

2.4 cm angle iron

2cm flat iron

2cm diameter pipe,105cm

bolts

2cm diameter pipe, 70cm

Source: Barker, 1990

Figure 21. Some tools and equipment designed by the Project for local manufacturer

150

strong cord. Thus the equipment was more dependent on practical ingenuity than sophistication. The materials were all widely available and cheap (Barker, 1990).

4.12 Economic analysis of the Project's small-farm system

Prior to the development of the Project's farm there had been no possibility of making a meaningful estimate of the likely economic viability of the small-farm system being proposed. There were no data. However, a cost-benefit analysis was recognised as an important priority so a preliminary estimate was made at the end of 1979 based on the results so far obtained (Bell and Dutton, 1979). The Project's farm system was complicated, over and above that on any small commercial farm, by having to undertake a range of tests and field trials, by having to breed as well as rear the livestock (crossing locals with exotics), and by having to monitor the results. The farm also carried considerable overheads associated with the presence of expatriates running the Project. For the purpose of the 1979 economic evaluation, therefore, it was assumed that the farm was owned and run by an Omani who already had a home on the land in which he lived with his family. For simplicity, we also assumed that the farmer worked with one type of animal, goats, that he had a herd of reasonable quality and that he was not involved in any sophisticated breeding work beyond the normal replacement of elderly breeding stock. Two alternative labour situations were examined: first, the work done entirely by members of the farmer's household; and second, the farmer employing 10 man hours of labour per day at the rate of RO100/month - conditions which prevailed on the Project's farm in 1978. In fact the Project employed relatively expensive Omani labour even though, as explained above, by then it was becoming more normal on the Batina for labour to be Pakistani for reasons of availability and cost. For simplicity in calculating, it was assumed that the paid labour worked entirely on the forage crops (irrigating and harvesting) and that all the crops were cut (as they mainly were in 1977 and 1978) although by 1978/9 we were beginning to appreciate the cost-cutting virtues of grazing.

It was also assumed that the animals kidded every eight months and that half of them produced twins. This high rate of productivity was attained at the second kidding of the Project's Omani goats in the winter of 1978/9. A 10% kid mortality was also assumed giving an overall productivity of two kids per doe per year. Finally, it was assumed that kids were sold at 7.5 months (32 weeks) and that breeding stock were replaced by purchase of mature animals locally.

The annualised fixed costs totalled just over RO2,000, labour accounting for over half the total (Table 13). The costs were based on the expenditure required to equip and cultivate the Project's 1.7ha farm. The same outlay was also assumed for a farm of one hectare. In the calculations only machines in regular use were accounted for - it was assumed that periodic cultivations would depend on hiring a locally-owned rotavator. The cost of operating the water pumps and the cutting machine (Table 14) are incorporated into the total Rhodes grass and alfalfa production costs (Tables 15 and 16).

Table 13. Fixed costs[1]

Item	Outlay	Cost/yr
(a) Water: two hand-dug wells and Impex GD 6.5hp pumpsets		
digging @ RO120 (10 years)	240	24.0
pumpsets @ RO220	440	88.0
installation @ RO50	100	20.0
spares/repairs @ RO50/set/yr	0	100.0
siphon hoses: 4 lengths @ RO18	72	14.4
plastic channel lining 500m @RO16/100m	80	80.0
(b) Cutting machine: 2-wheel cutter (Arun tractor)		
purchase @RO500	500	100
spares/repairs @ RO50/yr		50
(c) Labour:		
salary - 10/hr/day @ RO100/month		1,200
(d) Farm fence: chain link		
purchase and installation	1,500	300
(e) Buildings: shaded areas with chain link fencing		
4 rolls of chain link @ No 16	64	12.8
sheets of date fronds (*barusti*). 10 @ RO5	50	10
posts, 50 @ RO2	100	20
(f) Sand and gravel: flooring pens and paths		
5 loads @ RO8		40
(g) Electric fencing		
fencing unit (solar powered) 2 @ RO60	120	24
fencing posts	25	5
wire 10 of 200m roll @ RO2 (2yr)	20	10
Total	3,311	2,098.2
Total. without labour		898.2

Source: Bell and Dutton, 1979

(1) Arrangements for 1-1.7ha of crop. Amortisation over 5 years unless otherwise stated

Table 14. Machine running costs[1]

Item	Cost/hr (baisa)
(a) Water pump running costs	
diesel per pumpset, 4hr/l @ 48baisa/l	12.0
engine oil per pumpset, 70hr/l @ 440 baisa/l	6.3
Total	18.3
Total (2 pumpsets)	36.6
(b) Cutting machine	
petrol consumption, 4hr/l @ 54baisa/l	13.5
engine oil, 25hr/1.5l @ 440baisa/l	26.4
Total	39.9

Source: Bell and Dutton, 1979

(1) Assumed that only machines in regular use are purchased - i.e. rotavator is hired

Table 15. Rhodes grass - total costs[1]

Item	Outlay	Cost/yr
(a) Cultivation: assumed by hired rotavator		
3 passes, 16hr/ha @ RO1.2/hr	57.6	19.2
(b) Seed		
25kg of seed @ RO6.9kg	172.5	57.5
(c) Fertiliser		
compound (18.18.5) 100kg every 6 months; 200kg @		
RO3.75/50kg bag		15.0
ammonium nitrate (34.5%N), every cutting 100kg/ha - 22		
cuts/yr; 44 of 50kg bags @ RO3.75/bag		165.0
if grazing then nitrate application may be cut by half		(82.5)
(d) Irrigation (2 pumps: total output 10l/sec)[2]		
120 days @ 7 day intervals = 477hr		
245 days @ 10 day intervals = 681hr		
Total hours = 1158, @ 36.6baisa/hr		42.4
(e) Cutting[3]		
220 hours @ 40baisa/hr		8.8
Total	230.1	307.9
Total, (if grazed)	230.1	225.4

Source: Bell and Dutton, 1979

(1) Assumed that Rhodes grass is resown every 3 years. Costs and productivity calculated per hectare

(2) Irrigating every 7 days in summer and every 10 days in winter, depth 10cm - therefore 27.8hr/ha

(3) All cut - 22 cuts per year @ 10hr/ha

The total Rhodes grass production costs were then converted into cost per kg of green Rhodes assuming an annual yield of 152 t/ha/yr, the production of the Project's farm in 1978. The calculations also assumed, first, 1.0ha and second, 1.7ha, and that the latter entailed no additional fixed costs. On these assumptions the costs per kg (green) were:

1.0ha - excluding labour 5.16 baisa/kg

1.0ha - including labour 13.05 baisa/kg

1.7ha - excluding labour 3.87 baisa/kg

1.7ha - including labour 8.51 baisa/kg

The growing costs were far lower that the local sale price of alfalfa or Rhodes grass, and labour costs were shown to be a major cost element.

The alfalfa production costs were also converted into cost per kg of green alfalfa on the same basis as for Rhodes grass, above. On these assumptions the costs per kg green were:

1.0ha - excluding labour 6.53 baisa/kg

1.0ha - including labour 18.18 baisa/kg

1.7ha - excluding labour 4.63 baisa/kg

1.7ha - including labour 11.47 baisa/kg

Table 16. Alfalfa - total costs[1]

Item	Outlay	Cost/yr
(a) Cultivation: assumed by hired rotavator		
3 passes, 16hr/ha @ RO1.2/hr	57.6	19.2
(b) Seed: local seed from neighbouring farmers		
25kg of seed @ RO10.0/kg	250	83.3
(c) Fertiliser		
compound (18.18.5) 100kg at sowing; 100kg at first cut @		
RO3.75/50kg bag	15.0	5.0
superphosphate (45%N), @ 4.375/50kg bag (equal dressing every		
6 months)		17.5
if grazing then nitrate application may be cut by half		(82.5)
(d) Irrigation (2 pumps: total output 10l/sec)[2]		
120 days @ 4 day intervals = 834hr		
245 days @ 7 day intervals = 973hr		
Total hours = 1807, @ 36.6baisa/hr		66.1
(e) Cutting[3]		
160 hours @ 40baisa/hr		6.4
Total	322.6	197.5

Source: Bell and Dutton, 1979

(1)   Assumed that alfalfa is resown every 3 years. Costs and productivity calculated per hectare
(2)   Irrigating every 4 days in summer and every 7 days in winter, depth 10cm - therefore 27.8hr/ha
(3)   All cut - 16 cuts per year @ 10hr/ha

On the basis of the actual differences in production costs and yield figures obtained from the Project's farm in 1978, alfalfa was considerably more expensive to produce than Rhodes grass.

In order to cost the goats it was first assumed that the breeding females were fed a daily ration of 2kg each of alfalfa and Rhodes grass grown on the Project's farm, to which was added some concentrates at critical months in the pregnancy-lactation cycle. Finally it was assumed that the net cost of goat purchase (cost, mortality at 10%, cull value) was RO11 spread over 6 years, or RO1.83/head, and that veterinary costs were RO2.58/yr.

Based on rather different nutritional and health cost assumptions (concentrates, alfalfa, Rhodes and veterinary services), the costs of keeping two males for breeding were calculated, as were the costs of rearing the kids to 32 weeks of age.

From the Project's 1978 figures we assumed productivity per hectare of Rhodes at 152 tonnes/yr and alfalfa at 103 tonnes/yr. Given the feed requirements per 'breeding unit' of one female plus followers the farm should have been cropped in the ratio of 0.6ha Rhodes and 1.1ha alfalfa which would have provided sufficient forage for 76 breeding units, and two males. If these breeding units yielded 152 kids for sale at RO45 per kid (at 32 weeks) the net annual profit for the farm of 1.7ha would have been RO3 674 (assuming no labour cost), or RO2,504 (assuming paid labour).

Various alternative profit estimations were also calculated based on the following different assumptions: first, sales at 24, 32, 40 and 52 weeks of age;

and second, mean number of offspring per female per year of 1.35, 2.0 and 2.7 kids (all allowing for 10% mortality) (Table 17). At 2.7 kids/yr the does would have attained the theoretical 'maximum' - unattainable in practice - of 2 kids per litter every eight months.

The same feed and other costs per head were assumed, but of course if there were fewer kids per female or if they were kept on the farm for a shorter time before sale they would eat less forage and so it would be possible to keep more 'breeding units' of females plus followers.

Table 17. Total net profit, based on different sale prices and ages[1,2]

(a) Mean of 1.35 kids/yr							
	RO sale price						
Weeks	60	55	50	45	40	35	30
24	3,649	2,939	2,651	1,519	809	99	-611
32	2,810	2,200	1,590	980	370	-240	-850
40	2,188	1,648	1,108	568	28	-512	-1,052
52	1,498	1,038	568	118	-342	-802	-1,262
(a) Mean of 2.0 kids/yr							
	RO sale price						
Weeks	60	55	50	45	40	35	30
24	6,481	521	4,561	3,601	2,641	1,681	721
32	4,854	4,084	3,314	2,544	1,774	1,004	234
40	3,742	3,092	2,442	1,792	1,142	492	-158
52	2,618	2,088	1,558	1.028	498	-32	-562
(a) Mean of 2.7 kids/yr							
	RO sale price						
Weeks	60	55	50	45	40	35	30
24	8,825	7,690	6,555	5,420	4,285	3,150	2,015
32	6,426	5,536	4,646	3,756	2,866	1,976	1,086
40	4,940	4,200	3,460	2,720	1.980	1,240	500
52	3,497	2,902	2,307	1,712	117	522	-73

Source: Bell and Dutton, 1979

(1)   After deducting RO1,200 for labour and RO422 for fixed costs
(2)   Underlined figures are profit margins based on actual sales from the project farm, to Oct 1979

Table 17 clearly illustrates the importance of maximising the number of kids per doe per year, and minimising the number of weeks for the kid to reach a saleable weight.

The figures underlined in Table 17 are based on actual sales up to October 1979, and are net of labour (RO1,200) and fixed costs (RO422). They were helpfully indicative but it was recognised that much more long-term evidence of costs and sales of livestock would be required before firm conclusions could be reached about the viability of a goat (or sheep) small-farm enterprise.

The main point to stand out from the Project's kid and lamb sales of 1978 and 1979 (Figure 22) is the extreme range of price per kilogram: the mean for the five highest priced animals (RO2.3/kg) was 53% higher than the mean for the six lowest priced animals (RO1.5/kg). Figure 22 also indicates that the smaller animals fetched the highest price per kilogram, suggesting that highest profits would be made from animals which were small but had grown rapidly and were sold at a young age.

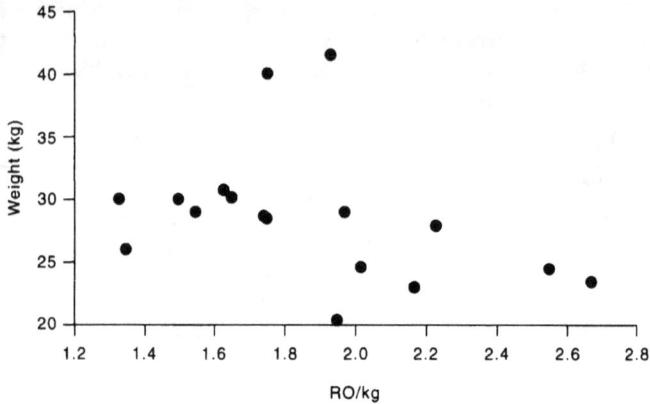

Source: Bell and Dutton, 1979

Figure 22. Sheep and goat lamb/kid sales for slaughter: weight against price/kg

On the question of singletons versus twins, Table 17 shows that when animals were sold at roughly equal prices per kilogram singleton animals would have had to show a 35-45% increase in growth over the twins in order to secure a greater profit; a 40-week-old singleton would have had to sell for RO50-5 but the 40-week-old twin only RO35-40. But we have already seen that the market was very undiscerning, with price differences per kilogram fluctuating by more than 50%, and the smaller animals selling relatively better than the larger ones. Therefore twins were likely to be more profitable than singletons.

In summary, the above analysis suggests that a farmer would have been advised to aim for a high rate of kidding and twinning, scientifically feed his kids (and lambs) to ensure rapid growth and sell the youngstock when they were still well under a year old. This system would have had the further advantage of reducing mortality rates - kids and lambs having less time to suffer illness and death. It would also have meant fewer breeding stock (and correspondingly less capital outlay for them and fewer problems from them) and fewer males to serve the females.

If half a farmer's stock produced twins, on the above costs and sale price basis, he would have made a profit of some RO3,000 after deducting RO1,200 for labour. This would have been a very attractive proposition to the typical smallholder on the Batina coast in the early 1980s.

However, for the small farmer there would have remained the problem of finding capital to start the enterprise, particularly as he would have received no return on capital until into the second year of operation. Let us look at the cost figures, assuming a rapid start to the work and based on the costs given above:

Item	RO
Water	912
Cutter	500
Fence	1,500
Female stock 76@RO60	4,560
Male stock 2@RO60	120
Rhodes grass (0.62ha)	286
Alfalfa (1.08ha)	446
Labour	1,200
Other	379
Total	9,903
Excluding labour	8,703

If we assume the farmer borrowed all the money (except for the labour cost) at 10% he had to repay RO870 each year. If through a scheme of OBAF the payments were deferred by one year and if the farmer made a profit of RO3,000 (after deducting labour) he could have repaid the interest and still cleared RO2,100 net income, equivalent to 24% of the borrowed capital. An alternative form of OBAF subsidy would have been to provide the farmer with improved stock at subsidised prices (or free) but leave him with the full responsibility for his stock thereafter. Such a scheme would also have created an opportunity for OBAF to improve mean stock quality. Training for the farmers could have been provided at al-Khabura.

The largest cost items were labour (RO1,200), concentrates (RO1,266) and crop production. The latter, including the cost of water and the cutting machine, cost RO880, and including labour, RO2,080. The number of stock is limited by the production of forage crops therefore every effort has to be made to maximise crop production - for example, increasing nitrogen on Rhodes grass for at least as long as the yield response shows a linear improvement, and minimising the ground lost to channels and ridges. The analysis also shows the importance of not overfeeding the stock - a 10% cut in the fodder ration would result in an increase of breeding females from 76 to 85, producing 18 extra kids. This would increase net income by RO600. But most important is labour. If in reality the farm system can be simple enough to operate so that all the work on a 2ha farm can be done by the family, then the net income to the family increases by a very high percentage. Therefore labour-saving ideas should be given very serious consideration, including maintaining an irrigation rate of at least 10 litre/sec, and grazing the stock instead of cutting the forage crops. With simplicity and time saving in mind, eliminating alfalfa and growing only Rhodes grass would have been beneficial even if it meant an additional requirement for concentrates.

Table 18. Livestock production levels at the Project during 1987 and 1988

	1987		1988	
Production indices	Number	%	Number	%
GOATS				
Adult does on farm[1]	72		72	
Kiddings	74		70	
KIDDING RATE[2]		102.78		97.22
Kids born	132		121	
Mean litter size	1.78		1.73	
REPRODUCTIVE INDEX[3]		183.33		168.06
Goats sold	58		137	
Goats died	64		25	
BALANCE	10		-41	
SHEEP				
Adult ewes on farm[1]	65		57	
Lambs	72		56	
LAMBING RATE[2]		110.77		98.25
Lambs born	114		86	
Mean litter size	1.58		1.54	
REPRODUCTIVE INDEX[3]		175.38		150.88
Sheep sold	96		147	
Sheep died	28		14	
BALANCE	-10		-75	
OVERALL				
Adult females on farm[1]	137		129	
Births	146		126	
BIRTH RATE[2]		106.57		97.67
Kids/lambs born	246		207	
Mean litter size	1.68		1.64	
REPRODUCTIVE INDEX[3]		179.56		160.47
Stock sold	154		284	
Stock died	92		39	
BALANCE	0		-116	

Source: Taylor, 1991e

(1) Average for each year (total nos of animal days as adult, for each female present, divided by 365
(2) Nos of births/100 adult females
(3) Nos of offspring/100 adult females

This was the position at the end of the 1970s. We were fortunate to be able to make a comparative economic analysis almost a decade later when Taylor (1991e) made a study of the final two full years of the Project's farm at al-Khabura, 1987 and 1988 based on its actual costs and income. At that time it remained true that our objectives had been pushed nearer to 'research' and a little further from 'development' than originally intended so that the management systems adopted were more varied and complex than those that a commercial approach would have used. The Project in consequence had retained its high and costly labour requirement of local staff and trainees, in addition to the complement of expatriate staff. Income was also sub-optimal because the staff were never under pressure to maximise the prices obtained from animals sold for slaughter or breeding. Nevertheless, Taylor's economic

158

analysis serves as a useful comparison (in the light of a decade of additional experience) with the one made earlier by Bell and Dutton (1979). It also highlights important areas for cost saving and indicates key areas for monitoring to help ensure that the best results possible are obtained, long-term, from the goat and sheep multiplication and development projects that MAF was initiating with leader farmers throughout northern Oman.

Table 18 summarises the Project's livestock productivity during the years 1987 and 1988. The lambing and kidding rates (births per 100 adult females) were not particularly high but the reproductive index (offspring per 100 females) was more impressive because of relatively high twinning ratios. The large number of goat deaths in 1987 was due to a unique outbreak of PPR.

Table 19 gives a breakdown of the goat and sheep production costs in both years. The area under forage was approximately 1.2ha (some land had been lost to a new al-Khabura road) of which about one-third was grazed by the stock and two-thirds was cut and stall fed or conserved as hay for winter feeding. By 1987, virtually all the forage grown was Rhodes grass which was cut each day and the wet weight recorded. The yield of the grazed area is here crudely estimated as half the amount cut and stall fed. Using this estimate the yield per ha was 92.4 tonnes for 1987 and 110.6 tonnes for 1988. These figures are of a similar order to those given in Dutton (1991), but lower than those obtained in 1978 (Bell and Dutton, 1979). Costs of cultivation, crop production, irrigation and harvesting had all risen between 1978 and 1987/8, so the variable costs of growing Rhodes grass on the Project's farm had increased from RO308/ha to RO415/ha. If the mean yield was 100 tonnes/ha (green) the cost per tonne was RO4.15. This compares closely with actual early 1990s practice at Wadi Quriyat (Ch. 5.2), where the farm manager estimated the cost at RO4.67 per tonne wet. For the sake of simplicity a nominal estimate of RO5/tonne wet weight was used in Table 19. On the Project other major feedcost elements included bought forage and concentrates.

The pen construction costs are those calculated by Winstanley and Hillman (1984), at RO1,100 for a 90 adult unit. Thus for the 160 adults in 1987/8 a figure of RO2,000 is used, depreciated over 10 years. Fixed irrigation costs are based on those calculated by Dutton (1991) at RO1,752 per 0.35ha. Depreciation is again estimated over 10 years. It is assumed that hired equipment would be used for cultivation of the field (so this is included in the forage cost) but that the farm would need to own one forage cutter, here given a nominal price of RO500, with an assumed life of five years.

On the above bases and assumptions, total costs were about RO31,000 in 1987 and RO29,000 in 1988, of which almost half are accounted for by the cost of feed.

Table 19. Livestock production costs, 1987 and 1988

Cost	1987			1988		
item	tonnes	RO/ton.	RO	tonnes	RO/ton.	RO
CONCENTRATES						
Total used	85.22			80.50		
Mean price/tonne		88.50			88.50	
Cost sub-total			7,542.0			7,124.0
Dates			385.0			212.0
BOUGHT FORAGE[1]						
Total used (wet wt)	100.63			99.10		
Mean price/tonne		57.22			55.65	
Cost sub-total			5,757.8			5,515.2
GROWN FORAGE						
Wet wt cut	73.9			88.46		
Wet wt grazed (estimate)	36.95			44.23		
Est. variable costs/tonne		5.00			5.00	
Var. cost sub-total			554.25			663.45
(Total forage fed)	(211.47)			(231.79)		
Total Feed Cost			14,239.05			13,514.60
OTHER VARIABLE COSTS						
Drugs[2] and veterinary			1,082.00			583.00
Vehicle hire (to market stock or carry feed)			1,089.00			151.00
Routine maintenance			290.00			429.00
Total Other Variable Costs			2,461.00			1,163.00
FIXED COSTS						
1 forage labourer salary[3]			1,860.00			1,860.00
7 stockmen salaries[3]			11,760.00			11,760.00
Depreciation on plant:						
Pens @ 2000/10 yrs			200.00			200.00
Irrigation systems @ 6000/10 yrs			600.00			600.00
Grasscutter @ 500/5 years			100.00			100.00
Total Fixed Costs			14,520.00			14,520.00
Total All Costs			31,220.05			29,197.60

Source: Taylor, 1991e

(1) All forage was recorded as wet weight
(2) Drugs were used to synchronise oestrus, which accounts for the major part of this item
(3) All Omani men

Table 20 summarises the total receipts. The sale prices of the young stock (3-15 months) seem low at RO48 for goats and RO44.5 for sheep. For comparison, a decade earlier in 1978/9 the farm had sold 10 cross-bred male goats at a mean price of RO49.2 and ten local male sheep at RO55.0. Also, in 1991 sales of male youngstock goats from the GMDP farms attracted a mean price of RO66.4 outside the *Id* periods, and RO75.5 at the *Ids* (Conroy, 1992b); females fetched RO67.3 and RO55.0, respectively, at these times. It is almost certain, therefore, that the Project's youngstock were undervalued when sold.

Table 20. Receipts, 1987 and 1988

Receipt	1987			1988		
item	Av. price	No	RO	Av. price	No	RO
STOCK SOLD						
GOATS						
Kids				18.75	4	75.00
Youngstock	48.03	40	1,921.20	36.26	81	2,937.00
Adults	66.00	18	1,188.00	63.03	52	3,277.56
SHEEP						
Lambs	29.00	2	58.00	19.04	23	437.92
Youngstock	44.55	85	3,786.75	45.47	100	4,547.00
Adults	55.50	9	499.50	46.60	24	1,118.40
Total Stock Sales		154	7,453.45		284	12,392.94
CHANGE IN VALUE OF STOCK INVENTORY[1]						
GOATS						
Kids	20.00	-10	-200	18.75	-7	-131.25
Youngstock	48.03	18	864.54	36.26	-22	-797.72
Adults	66.00	2	132.00	63.03	-12	-756.36
SHEEP						
Lambs	29.00	-4	-116.00	19.04	-32	-609.28
Youngstock	44.55	-7	-311.85	45.47	-26	-1,182.22
Adults	55.50	1	55.50	46.60	-17	-792.20
Total Value Change		0	424.19		-116	-4,269.03
OTHER INCOME						
Sale of compost/manure			185.20			68.50
Sale of wool/hair						5.00
Total Other Income			185.20			73.50
Total Income			8,062.84			8,197.41

Source: Taylor, 1991e

(1)   (Kids = < 3 months; youngstock = 3-15 months

Table 21. Margins, 1987 and 1988

	1987		1988	
Items	Stock	RO	Stock	RO
Number of adult females	137		129	
Total income		8,062.84		8,197.41
Feed costs		14,239.05		13,514.60
Margin Over Feed		-6,176.21		-5,317.19
Margin Over Feed/adult female		-45.08		-41.22
Other variable costs		2,461.00		1,163.00
Gross Margin		-8,637.21		-6,480.19
Gross Margin/adult female		-63.05		-50.23
Fixed costs		1,452.00		1,452.00
Net Margin		-23,157.21		-21,000.19
Net Margin/adult female		-169.03		-162.79

Source: Taylor, 1991e

Table 21 shows the margins. It is clear that the Project's enterprise did not make money. But can conclusions be drawn from the experience of this research orientated farm that can usefully be applied to the local small farmers, and particularly to the GMDP farms? One difference was that the Khabura Project had high drug, labour and overhead costs not applicable to small, owner-operated farms.

High feed costs, however, are common to all livestock enterprises, but they were particularly high at the Project's farm making it worthwhile to recalculate feed costs assuming different feeding regimes. To make an enterprise financially sustainable the revenue must exceed feed costs with a margin that covers all other costs and leaves a reasonable return for the operator. Unfortunately, on the Project the margin over feed was substantially negative for both years analysed (Table 21). There were two major causes of this. First, half the forage requirement was purchased at very high cost (higher even than market prices for commercially grown Rhodes grass hay). Second, milled concentrates made up a high proportion of the total ration.

Table 22. Feed costs and margin over feed if the Project had grown all its forage

Cost item	1987			1988		
	tonnes	RO/ton.	RO	tonnes	RO/ton.	RO
CONCENTRATES						
Total used		88.50			88.50	
Mean price/tonne	85.22			80.50		
Cost sub-total			7,542.0			7.124.0
FORAGE[1]						
Ttl forage, assumed grown	211.47			231.79		
Est. variable costs/tonne		5.00			5.00	
Var. cost sub-total			1,057.35			1,158.95
Total Feed Cost			8,599.32			8.283.20
Total Income			8,062.84			8,197.4
Margin Over Feed			-536.48			-85.7
Margin Over Feed/adult female			-4.03			-0.6

Source: Taylor, 1991e

If the forage had all been home grown the margin over feed would have been as illustrated in Table 22. The feed cost would have been reduced to RO8,599 for 1987 and RO8,283 for 1988 making the margin over feed per adult female RO-4.03 and RO-0.65; a great improvement but still negative.

The high level of concentrates in the feed ration also contributed to the high overall feed cost. Working backwards from the figures for total forage and concentrates fed and the herd/flock composition (calculated from stock records during 1987 and 1988) the approximate rations fed to different classes of stock can be estimated. The estimate (Table 23), which is in agreement with the estimate verbally reported by the Project's staff, is that concentrates formed up to 80% of the daily dry matter intake for both young animals and production stock, though the mature stock on a maintenance diet were mainly reliant on wet forage - a mix of Rhodes grass and alfalfa.

Table 23. Rations fed to the Project's livestock as estimated from overall feed use

A. Approximate rations as fed (kg/day)

Ration category	Concentrates	Wet forage	Dry Matter (DM)
Young	0.93	1.05	1.14
Production stock	1.30	1.00	1.50
Maintenance stock	0.30	3.98	1.10

B. Nos. of animal days in stock and ration category, 1987 and 1988 combined

Stock category	Total animal days, 1987-8	Ration category		
		Young	Production[1]	Maintenance
Young	96,360	96,360		
Adult female	97,090		41,263	55,827
Adult male	19,710			19,710

Source: Taylor, 1991e

(1)   Production = last 2 months pregnancy + 3 months lactation
      (Yearly pregnancy rate: 102% (272 pregnancies from 133 females)

Table 24. Approximate rations and feed quantities, as suggested

Feed	Ration category			
	Young	Production	Maintenance	Total
Concs./day (kg)	0.5	0.3	-	
Total concs. (ton.)	48.1	12.4	-	60.5
Forage (wet)/day (kg)	3.0	6.0	5.5	
Total forage (ton.)	289.1	247.1	415.5	952.1

Source: Taylor, 1991e

Table 25. Feed costs and margin over feed - assuming forage all grown on farm

Ration item	RO/ton.	tonnes	RO
CONCENTRATES			
Total required	60.56		
Mean price/tonne		88.50	
Cost sub-total			5,359.47
GROWN FORAGE[1]			
Total required	952.11		
Ttl forage assumed grown	952.11		
Est. variable costs/tonne		5.00	
Var. cost sub-total			4,760.56
Total Feed Cost			10,120.03
Total Income			16,260.25
Yearly Margin Over Feed			3,070.11
Yearly Margin Over Feed/adult female[1]			23.08

Source: Taylor, 1991e

(1)   Assumes 133 adult females on farm

The concentrates were more expensive than the home-grown forage, therefore if forage had substituted for concentrates to a greater degree the feed costs would have been cheaper. The suggested rations, Table 24, are similar to those being fed at ABARC and have the same dry matter intake as the actual diet at al-

163

Khabura. It is here assumed that they would have maintained the same level of kid/lamb production and growth.

According to the feeding regime suggested by Taylor, 60.5 tonnes of concentrates and 953 tonnes of wet forage would have been required (Table 24).

By substituting home-grown forage for some concentrates the Project's livestock enterprise would have returned RO23 over feed per adult female, as shown in Table 25.

Table 26. Receipts, 1987 and 1988 - assuming higher sale prices and lower mortalities

Receipt item	1987			1988		
	Av. price	No	RO	Av. price	No	RO
STOCK SOLD						
GOATS						
Kids				20.63	4	82.5
Youngstock	52.83	80	4,226.6	39.89	91	3,629.6
Adults	72.60	18	1,306.8	69.33	52	3,605.3
SHEEP						
Lambs	31.90	2	63.8	20.94	23	481.7
Youngstock	49.01	95	4,655.5	50.02	100	5,001.7
Adults	61.05	9	549.5	51.26	24	1,230.2
Total Stock Sales		204	10,802.2		294	14,031.1
CHANGE IN VALUE OF STOCK INVENTORY[1]						
GOATS						
Kids	20.00	-10	-200	20.63	-7	-144.4
Youngstock	52.83	18	951.0	39.89	-22	-877.5
Adults	72.60	2	145.2	69.33	-12	-832.0
SHEEP						
Lambs	31.90	-4	-127.6	20.94	-32	-670.2
Youngstock	49.01	-7	-343.0	50.02	-26	-1,300.4
Adults	61.05	1	61.1	51.28	-17	-871.4
Total Value Change		0	489.6		-116	-4,695.9
OTHER INCOME						
Sale of compost/manure			185.2			68.5
Sale of wool/hair						5.0
Total Other Income			185.2			73.5
Total Income			11,474.0			8,197.4
Total Income for Two Years						20,882.6
Total Feed Cost						10,120.0
Yearly Margin Over Feed						5,381.3
Yearly Margin Over Feed/adult female						40.46

Source: Taylor, 1991e

(1) Kids = < 3 months; youngstock = 3-15 months

Additionally, other factors specific to the Project's farm may have made the return lower than could have been achieved on a local small farm. First, the prices received for stock sold were lower than the true market price, partly due to cross-bred animals being less favoured in the market, and partly due to a lack of time/incentive by the Project's staff to maximise selling prices, as shown above. Second, mortality at the Project's farm was higher than would be expected on a small farm using only local stock, and 1987 was an unfortunate year for the Project due to a unique outbreak of PPR causing 30% mortality in goats less than one year old.

Table 26 shows the effect on margins of a 10% increase in goat and sheep sale prices and an extra 80 stock sold over two years (if PPR had not happened and other mortalities had been slightly less), and assuming the lowest feed costs.

According to the table a yearly margin over feed of RO40 per adult female could have been achieved. But important assumptions are the use of PMSG to increase litter size, the cost per tonne of home grown Rhodes grass and the ability to maintain production levels with a low dependency on concentrates. The experience of the Desert Agriculture Project (DAP) farm in Marmul, Southern Oman (Weber, 1989) seems to confirm that the concentrate levels proposed in Table 25 are appropriate. On the DAP farm the goats were fed entirely on good-quality Rhodes grass, with no concentrates at all. The breeding females suffered only slight weight loss during lactation and the kids growth rates were only slightly lower than those at al-Khabura. But for a small farmer it is important to note that if, say, he has 30 does and a reproductive index as high as the Khabura Project's farm, his total margin would be RO1,200 which would only be an attractive return if he did not use any paid staff. GMDP farms are only likely to be viable if owner-operated.

In summary, these two economic analyses of 1978/9 and 1987/8 are in agreement on several important points: the importance of the high cost of labour; the need to maximise forage production on-farm; the need to maximise the twinning rate and reduce the length of the breeding interval; and the importance of minimising mortalities. However, Taylor's studies also show that the breeding rate did not maintain the 1978 aim of 2 kids/lambs annually. Nevertheless, it was possible to envisage a successful small to medium-sized unit with 30-50 does providing meat, milk and a useful income but as part of a family enterprise including other income sources.

4.13 Extension and training

Extension was a major component of the Khabura Project's work from the outset. The development of intensive systems for the production of sheep and goats in northern Oman could only have been justified if they were adopted by the farmers. Extension was the vital link between the Project and the farmers to ensure that the goat and sheep production systems were understood and adopted by those farmers interested in improving the quality and output of their livestock units.

The Project's field trials and demonstration activities have already been discussed, including those that took place on the Project's own farm ('on-station') and on farms belonging to some of the people with whom we worked most closely (here called 'on-farm' trials and demonstrations).   The demonstrations always contained elements of training and extension. These and other aspects and approaches to training and extension as planned and as executed - not always the same thing - are discussed below. Initially most of the trials and demonstrations were conducted on-station in order to prove to ourselves (as much as to the farmers and other rural producers) that, technically, they were feasible and with a real economic potential. This work absorbed a lot of our time and so, in the early years, on-farm extension was the work of one member of the team only.

January 1986 saw the change of Project sponsorship from PDO to MAF. Discussions with MAF livestock officials had already shown that the Ministry was interested in us expanding our training and extension role, within the context of its own field programmes. Also, by this time there was a growing confidence that our evolving livestock system was appropriate for small-farmer requirements. So, the time had come to reshape and expand the Project's extension inputs. The long time-span of the new agreement with MAF, 1986-90, together with the expansion of the number of specialist Project staff from four to five, had created the opportunity for a radical review of extension methods. Under consideration were, the creation of an extension team, working with groups of farmers, and the expansion of the technician training programme for employees and secondees to be achieved in part through closer working links with other organisations.

4.13.1 One-on-one extension contacts
When the Project's on-station field trials were well established and had been demonstrated to a growing number of people, and their value had been confirmed by analysis of the Project's records, it was then feasible to devote a growing proportion of staff time to extension. Initially almost all the extension was conducted on a one-to-one basis. Individual farmers either came onto the Project's farm for advice or else Project staff made visits to local stock owners and other farmers. It was important to start with one-to-one extension approaches because the Project was still learning a great deal from the farmers about how best to adapt its livestock and irrigation systems to the realities of small-farmer social and economic circumstances. Also, it was best to work with individuals if that produced a few fully satisfied clients who themselves acted as secondary extension agents in everyday conversation with their friends. Working with larger groups of semi-satisfied clients could have had a long-term damping effect on progress.

The initial work was undertaken by topic. The Project's livestock system and the irrigation system broke down naturally into a large number of component parts. Initially, all the local farmers and stock owners who approached the Project were primarily interested in one or other of these components. Some wanted information about enlarging their irrigation basins, improving their water channels, using the rotavator or growing Rhodes grass. Others were primarily interested in better animal nutrition, obtaining higher-quality livestock, building better animal pens, obtaining improved veterinary support or making fuller use of animal by-products. Yet others were concerned with water-efficient irrigation systems and irrigating fruit trees and vegetables.

The Project increasingly developed one-on-one contacts by conducting trials and demonstrations on farmers' land or with farmers' livestock. These activities have already been described. They included:

- the early trials crossing farmers' does with the Project's Anglo-Nubian bucks;
- finding potential veterinary auxiliaries (paravets), training them and running a livestock clinic for local stock owners;
- testing the concept of a village milking parlour and dairy for goats, and stocking and equipping small dairy goat herds;
- revitalising the spinning and weaving industry;
- marketing composted goat and sheep dung;
- using rotavators, and testing structures for their sale, maintenance and hire;
- growing Rhodes grass, including inter-cropping in palm gardens;
- the local manufacture and sale of fibre-reinforced-cement irrigation channels;
- trials and demonstrations of low-head, low-technology trickle irrigation systems for fruit and row crops;
- the design and construction of low-cost goat and sheep pens;
- the design and local manufacture of farm tools and equipment.

Between them these activities brought the Project into close working relationships with hundreds of farmers and craft-scale manufacturers, almost all with a direct or indirect interest in livestock husbandry.

Naturally enough some stock owners became interested in more than one component of the Project's livestock system but, in the early years, no one was prepared to adopt a version of the system in its entirety. This was not surprising. People were taking their time to acquire confidence in the new range of techniques. The Project did not try to hurry them, again taking the opportunity of learning from the farmers how best to adapt a particular part of the system to fit each farmer's needs.

However, it had become clear, by about 1984-5, that in addition to the hundreds of farmers who had adopted one or another of our individual tried and tested contributions to improved crop and livestock husbandry, a growing number were adopting a scaled-down version of the Project's livestock system as a whole. Almost all of them based their work on good quality cross-bred goats and sheep purchased from the Project. These farmers, who we called 'participant farmers', were about 60 in number by the end of 1985. By this time we had had to make some important decisions about how best to utilise the time of the senior Project staff. Already, in 1981, as has been described, we had handed over most of the responsibility for our veterinary clinic to a young Omani trainee partly because the then extension specialist did not have the time to treat the sick animals of the all the farmers with whom we were working. In 1984/5 we decided that we should focus future extension work primarily on the participant farmers because they represented our ultimate farm systems goal of integrated small-farm improvement, and because a single extension visit could cover a wide range of related activities. Thus the participant farmers received special attention from the Project's extension officer including regular monthly visits. This group of 60 farmers formed the base upon which we hoped extension and training progress would be built during the 1986-90 Plan period. It did not mean, however, that other farmers were neglected, but shortage of time meant that contact with them had to be curtailed. They were visited less frequently, as occasion demanded or because they were making valued contributions to off-station trials and demonstrations.

Table 27. Extension visits to participant farmers, 1986-7

1986		1987	
Time period	No of visits	Time period	No of visits
Jan-Mar	186	Jan-Mar	65
Apr-Jun	86	Apr-Jun	85
Jul-Sep	63	Jul-Sep	45
Oct-Dec	90	Oct-Dec	95
Total	425	Total	290

Source: Ann Reps. 1986 and 1987

In practice the desired monthly rate of extension visits to participant farmers was rarely achieved, as Table 27 illustrates. The low rate was in part due to shortage of staff because of delays in signing the new contract with MAF in 1986. Also, in the latter months of 1987, in response to an opportunity created by discussion with MAF, we were becoming more involved in training technicians in partnership with the Nizwa Agricultural Institute. Additionally, MAF by then was actively responding to the policy switch introduced by the new Minister of Agriculture who wanted us to take a more national role and, amongst other things, help cater for the needs of a new group of livestock 'leader farmers' to be selected throughout northern Oman. The required range of support activities for the leader farmers was very similar to that which we were providing to our participant farmers. In the al-Khabura area 12 of the latter became leader farmers within the national scheme. Therefore, locally we concentrated our integrated extension approach on these 12 farmers, and initiated our own monitoring procedures on their progress. By October 1988, after the arrival of Iain Rogerson as our training specialist in September, the 12 GMDP farms were being visited regularly. Work was also in hand to select another 12, a good proportion of them being farms belonging to the Project's own workers.

4.13.2 Setting up an extension team
The plan for the 1986-90 phase of the Project was that extension would become a team responsibility (Dutton, 1987c). It was anticipated that this would generate more ideas, integrate all aspects of the husbandry systems that were being demonstrated, have a stronger impact, ensure continuity and allow time for the development of working links with other organisations. It was envisaged

that the extension team would work with groups of farmers, perhaps establishing a farmers' association with which it would co-ordinate its activities.

4.13.3 Working with groups of farmers

Up until the end of 1985 most experience of working with groups of local producers had been obtained through the spinning and weaving project. This worked in two ways. In some cases groups of spinners/weavers (all women) had been brought into the Project's weaving workshop to learn a new technique. More typically, however, the weaving adviser made visits out to the villages in which the women lived - their looms were in their homes. Later, women gathered together in one of the houses and learned not only from the weaving adviser but also from each other (Ch. 4.4). Then, when the weaving project opened its branch in Wadi Sahtan (near Rustaq) women from that region came for instruction (and for more informal learning from each other's experience) to a shaded classroom established in the village of Ayn Amq.

By 1986 it was possible to envisage a range of livestock production activities operating in a similar manner. The farming community around al-Khabura and MAF approved our approach to the development of small-farm livestock systems. So, with sponsorship now coming from MAF, there was a need to bring information about the Project's work to the attention to a much wider range of farmers and other related rural producers, and to achieve this goal with all reasonable speed. It was appropriate, therefore, to work with groups of farmers where possible. It was hoped that formal or informal groups of farmers (and other rural producers) would come together in order to make outside visits or to come onto the Project's farm to participate in field demonstrations, to watch audio-visual presentations, to listen to speakers or to contribute to discussion meetings.

In fact, in 1986 the beginnings of a farmer's group already existed, composed of those 60 or so 'participant farmers' referred to above. These farmers shared certain characteristics. Most of them had known and worked with the Project's staff for several years. They often visited the Project's farm to purchase livestock, seeds, concentrates, etc. or to bring in sick animals for treatment or to discuss their work. Project staff made regular visits to their farms. These participant farmers had mostly demonstrated, over a long period of time, an above average level of skill and interest in sheep and goat husbandry combined with a desire to improve their knowledge, ability, facilities and stock quality. Most of the participant farmers had adopted a scaled-down version of the Project's sheep and goat husbandry system. On the forage side they hired the Project's rotor tiller, grew Rhodes grass, used nitrogen and phosphate fertilisers and had enlarged and rationalised the layout of their irrigation basins. As for livestock, they had purchased from the Project improved, cross-bred sheep and goats which they were keeping in well-constructed pens and feeding a better diet. Some of the livestock were for sale or for eating, others, particularly the good females, were being kept for breeding. The animal dung was being cleaned from the pens and dumped into pits where it was being rotted down into compost. The animals, with assistance from the Project's paravets, were being maintained in good health.

We envisaged forming, from the group of participant farmers, a farming organisation to be called, perhaps, the al-Khabura Stockfarmers' Association. Also included in such an organisation would be those farmers keeping goats or sheep for milk. An alternative name, if the association included those farmers who were equally responsive to the Project's ideas, but with a greater interest in modernising their forage irrigation systems and growing fruits and vegetables than in keeping livestock, might have been the al-Khabura Farmers' Association.

It seemed most appropriate that the Project's farm should remain (into the then foreseeable future) the best site for at least some field demonstrations of techniques and equipment, including the design and construction of pens, the nutrition of kids and lambs, standard veterinary procedures, dairy work, operation and maintenance of machinery, forage crops and irrigation systems. We also discussed the possibility of making audio-visual aids (with help from Sultan Qaboos University's audio-visual centre) and arranging a programme of visiting specialists. Visits to other locations were also very much on the agenda, including: local farms, modern farms (including those belonging to MAF, Sultan Qaboos University (SQU), the Nizwa Agricultural Institute and a modern farm in the private sector), and other organisations such as OBAF and and the Public Authority for Marketing Agricultural Produce (PAMAP).

4.13.4 Training employees and secondees

By 1984 a total of 24 people, mainly Project employees, had acquired new agricultural and craft skills (Table 28). Some of these men were certainly able enough to train other men. Their total does not include the very large number of farmers whose skill level had been upgraded as a result of the many on-farm trials listed above. Nor does it include the many women who were actively participating in the spinning and weaving project.

Table 28. Training in practical skills on the Project

Skill	Number
Power tiller operator	4
Power tiller mechanic	1
Diary technician	1
Craft/weaving workshop technician	1
Forage harvester operator	2
Paravet	2
Water channel manufacturers and installers	5
Trickle irrigation system installers	2
Livestock assistants	3
Forage crop assistants	2
Extension assistant	1
Total who had acquired new skills	24

Source: Dutton, 1984

The Khabura Project had always employed, amongst its Omani staff, a mixture of trainees and labourers. Amongst the former there were a number of young to middle-aged men and one woman. Typically they were active, intelligent people with practical aptitudes and interests but without much formal education. Such people, we fully recognised, were vital to the success of any agro-rural development programme. When the Khabura Project became part of MAF it had the opportunity, and duty, to build on its technical training experience with these people. The Project was in a position to train or retrain staff from MAF agricultural extension centres from all parts of Oman. In discussion with other organisations we aimed to undertake an expanded training role in close co-operation with various branches of MAF and with Sultan Qaboos University. The training, it was widely agreed, could then take place at more than one site, thereby enhancing the trainees' experience and creating the opportunity of including both practical and theoretical components. It was anticipated that courses for the trainees would be of the 'in-service' type. In addition, it was also agreed that the courses should be properly structured and formally recognised so that their successful completion resulted in an increase in the trainees' subsequent level of responsibility and salary when they were employed in a technical capacity.

During the period 1986-90, for all 11 skills listed in Table 28, it was felt that the Khabura Project should play a training role of growing importance. The envisaged role, in each case, was to teach practically orientated field skills to people who would subsequently be employed to train farmers (and other rural workers) or to train trainers.

We believed that our most valuable training contribution would be for:

- irrigation and water mangement (trickle, channel and sprinkler delivery systems);
- field crop cultivation, husbandry and harvesting (forage crops);
- livestock husbandry, especially for frequent breeding systems;
- dairy work;
- paraveterinary skills;
- operation and maintenance of power tillers.

The future Khabura Project role in training paravets was already the subject of on-going discussion with MAF as a result of our previous success in paravet training. At a meeting in the Department of Livestock Wealth on July 8th 1984 it was agreed that:

'Omani nationals could be trained in the paraveterinary area as was evidenced by the success of the assistants at al-Khabura who are well accepted by farmers', 'it was unanimously agreed that a training programme was essential and the al-Khabura Project did have a significant place to play in an overall programme', 'the type of training envisaged would produce assistants capable of being employed in either veterinary clinics, quarantine services or on research stations', 'al-Khabura could provide practical training in routine veterinary treatments, exposure to animal husbandry practices of goats and sheep under intensive management conditions and experience with the 100+ farmers the Project was regularly involved with', and 'it would be necessary under current staffing levels to provide extra accommodation or rent quarters for trainees [at al-Khabura]'.

It was also stated that based on rented accommodation at Rumais together with the utilisation of the Khabura Project facilities, 200 recruits could be trained over a 5-year period at a cost of RO5,000 per person - a tiny fraction of the cost of full veterinary training overseas.

Also in 1984, discussions were initiated with MAF about incorporating the power tiller work at al-Khabura into the Ministry's national farm machinery scheme, operated with loans and subsidy from OBAF. By 1985 these discussions had resulted in the following: the transfer of a power tiller and trailer to al-Khabura; moves towards the selection of two trained tractor drivers for re-training at al-Khabura as power tiller instructors (to train farmers) and an invitation to power tiller dealers to introduce their products to trainees at al-Khabura and provide manuals, instruction, parts and special product training for operation, maintenance and basic repair. According to the minute of a meeting held in MAF on 23rd June 1984 it was agreed that:

'The trained instructors will be directly under the supervision of the Khabura Project and the farmers' training programme will be established jointly by both the training section of the Ministry and the Khabura Development Project.'

It was anticipated that according to MAF policy the number of farmers owning, operating and privately hiring out power tillers in the region would grow steadily in the 1986-90 period and that, therefore, there would be a growing need to adapt and develop servicing and maintenance and mechanic training facilities for a programme that would expand the length of the Batina coast. It was also envisaged that instructors would be trained for other parts of Oman and that demonstrations could be given in other centres, such as Rustaq.

In mid-1987, for veterinary assistants, the Khabura Project and MAF staff discussed the feasibility of establishing practical demonstration and training courses at al-Khabura, to be taught in conjunction with the theoretical courses being run at the Nizwa Agricultural Institute. At Nizwa 36 MAF employees, working in different regions of Northern Oman, were being given in-service theoretical veterinary training. It was agreed that these students, in nine batches of four, should each spend a fortnight at al-Khabura where they would undergo an intensive, hands-on 10-day training session.

By the end of December, 15 men had attended these in-service training courses, learning paraveterinary and livestock extension skills and 'seeing cases'. The courses at al-Khabura were severely practical and were able to rely on an abundant supply of veterinary cases arriving at the extension centre, to which the Project's clinic had now moved. Each day followed a similar schedule. In the early morning the trainees observed the Project's livestock, learning to identify veterinary problems at an early stage. Subsequently the trainees treated the stock under the supervision of the Project's veterinarian - this encouraged the practice of routine and systematic observation of stock for health problems and it also taught about ear tagging systems and animal identification. Next, the trainees attended the government clinic, under the supervision of the veterinarian, from 09.00 to 10.30, during which time they were given as much practical work under supervision as possible. From 10.30 to midday the trainees made extension visits to the 'participant farmers' where animals were observed and treated for problems, or routine interventions were conducted such as vaccination or anthelmintic dosing. From 14.00 to 17.00 trainees were at the clinic, or continued with extension activities with participant farmers. One morning of each course was set aside for introductory talks on other aspects of the Project's work, particularly irrigation and goat dairying. The courses continued into the spring of 1988 by which time over 30 agents had attended. A report was submitted to MAF on each trainee, who were assessed according to four criteria: enthusiasm for obtaining practical experience, actual practical ability, communication ability (Omani to Omani), and co-operation, including punctuality.

Andrew Gauldie from mid-1988 became involved in assisting the UNDP veterinary volunteer, Dr Albuchair, with the design of a syllabus and training procedures for the new courses being established at Nizwa. Dr Albuchair started training at Nizwa in August 1988, continuing the admirable course that Dr Mustafa Khater had been teaching the previous year. At the same time we, in al-Khabura, were taking trainees from the field. They were the same men as before, again for a very practical course with an emphasis on hands-on experience and the appropriate use of visual aids.

On the basis of the 1987/8 experience we reshaped and extended the courses to include other related topics to be taught during the winter season 1988/9. By mid-1988 we had put a training proposal and schedule to MAF, to include:

- animal production and veterinary health;
- irrigation and fodder production;
- dairying;
- small-farm management;
- extension skills.

The courses were specially designed to fit in with MAF livestock and fodder production projects for small farms. Unfortunately, the proposal to teach dairy work as a specialist course foundered because no men (and only men were available) expressed any interest in it. However, some background training in dairy principles was included in the Animal Production and Health course. Nick Foster had also received the go-ahead from MAF to train, at al-Khabura, up to 12 men from the Batina extension offices on a full irrigation course - the same as the one being delivered at the Nizwa Institute for staff from the interior extension centres.

The courses were grouped into four batches, in successive summers and winters, and included a farm show at the end of each year.

The primary objective of the courses was to enhance the practical skills of the government's cadre of extension agents, particularly those concerned with the MAF livestock and forage production programmes and especially with the GMDP (and its proposed sheep counterpart). The courses were all very practical in nature, and designed to demonstrate to the agents both up-to-date techniques and the use of modern equipment (Gauldie, 1988). We recommended that the practical courses should be held in conjunction with theoretical courses to be held either at Rumais or Nizwa. Each course would be led by one of the five specialists then stationed at al-Khabura: General Extension Skills (Iain Rogerson), Small-Farm Management (Andy Barker), Veterinary Skills (Andrew Gauldie), Dairying (Kate Rogerson), Irrigation and Fodder Production (Nick Foster and Andy Barker) and Animal Production (Andy Barker), with Iain Rogerson, our training and extension specialist, as overall co-ordinator. The team of specialists were to provide close, individual supervision for the trainees.

Each training programme divided into two parts, the first in the period September to November 1988, and the second in the period January to March 1989. This created the opportunity for each extension agent to attend one of the three specialist courses and, in the second phase, a more broadly based Extension Methods and Small-Farm Management course.

Up to 42 agents could be catered for during the first year, with the majority of the places allocated to the livestock course, to accord with perceived demand.

A range of other courses was also designed to be followed by the Project's own Omani staff, 'leader farmers' and young Omani professionals (such as graduates from SQU) who, we hoped, would provide future staff for the Khabura Project in its next phase, 1991-5.

Brian Addy, Livestock Co-ordinator in the Department of Animal Wealth (DAW) of MAF, believed that the Khabura Project had an important future role to play during 1991-5, i.e. in the period of the coming 5-year Plan. He envisaged it as a 'Rural Development Centre', such as he had experienced in Botswana, with a role of 'adaptive research' - converting ideas from the Ministry and the research information from Rumais into farming systems which were applicable to the needs and capabilities of small and middle-sized farmers. He also saw that the Project could further develop its training role. Indeed, it had occurred to us that the Project could in its entirety have become a training centre for agricultural technicians. Field technical staff were in short supply in Oman and without them it was difficult to envisage the farming community rising to the challenge that MAF had, for example, set them with the GMDP scheme. By November 1988 we were already into the second year of giving practically based courses at the Project to extension workers being sent to us by MAF from all over northern Oman. Both MAF and the students appeared to like what we were doing (Rogerson, 1989).

Addy (1989) emphasised that the GMDP needed increased support from trained extension staff, and studies to monitor and evaluate the project impact. This was in addition to the training of veterinary assistants and nurses that was being catered for by the UNDP supported programme at Nizwa with practical training being provided by the Project. There was also a requirement to train the newly recruited Animal Production Assistants. Though they were all graduates from Nizwa College they still needed further specialised training in disciplines related to their field responsibilities. This was being planned also to include practical inputs at the Khabura Project. He stressed that a broad consequence of the GMDP was that:

'The need for a facility for in-service short courses in a range of animal health and production subjects is very apparent to allow for the updating of field staff in both technical skills and the details of Departmental projects, programmes and policies.'

A logical place to run these courses, particularly their practical aspects, was the Khabura Project which then had a very appropriate range of staff expertise coupled with its own small sheep and goat farm and excellent contacts with the local farming community. It was also situated in the midst of small and larger farms, in the middle of the Batina region, which is Oman's principal farming region. Trainees attending short courses at the Project would have had every opportunity to develop their practical skills under the guidance of expert tuition. The courses would have been an extension of those already initiated, and the facility could have been expanded to become a broad-based practical agriculture training facility, or, more generally, a Rural Development Centre to include also such things as financial management and training in small rural craft-industrial enterprise skills. Within this scheme it was hoped that Iain Rogerson would become the Technical Training Officer to the D-G of Agriculture with responsibility to identify, plan, design and deliver practical training courses, primarily (but not solely) in support of the GMDP.

4.14 Towards a new rural production system?

In summary, the Khabura Project had examined every aspect of a new rural production system based on goats and sheep, and come near to success in achieving its aims. Indeed, it could be said to have succeeded since some 60 local 'participant farmers' had adopted a version of the goat and sheep management system and, as will be shown below, the Minister of Agriculture had then asked us to play a central role in converting the sub-regional impact of the Khabura Project into a national goat/sheep breeding and 'leader farmer' scheme. Yet true success would only have been achieved if a higher proportion of the participant farmers in and around al-Khabura had achieved sustainable economic success and, in the process, created employment for manufacturers and providers of the goods and services they required and for processors and marketers of the meat and by-products they produced. This would have rebuilt, at a higher level of sustainable production, the mutual self-reliance and the productive interdependencies that had characterised Oman's rural communities in pre-oil days.

The people were certainly interested in the Project and hundreds of them adopted one or more of its ideas. Most were trying to improve an existing farm enterprise while others were learning skills that could have expanded into successful small businesses - running a village dairy, creating a paraveterinary practice, installing trickle irrigation systems, making irrigation channels, operating rotavator services, composting and selling goat and sheep dung, making and repairing farm tools and equipment, and spinning and weaving sheep's wool and goat hair.

However, the key issue was the profitability of the goat/sheep small-farm system and associated enterprises. At a time of immensely rapid and therefore very unsettling rural change, when the 'old' knowledge associated with date palm production and extensive livestock husbandry was no longer of economic value, the government adopted a policy of cheap food imports and therefore removed much of the incentive for learning new production systems. The mass importation of cheap Australian sheep limited the price rises of local livestock at a time when farming needed to be very profitable in order to encourage active young men to stay in the industry and learn new skills. Similarly, the highly subsidised services provided through the extension centres inhibited entrepreneurial initiative for the provision of rotavator, seed, fertiliser and veterinary services. Also, the equally subsidised provision of imported piped irrigation systems inhibited the local development of assembly, installation and maintenance skills. Furthermore, government bureaucracy introduced rules that crippled the dairy unit and prevented recognition of our highly skilled paravets, whilst divisions of responsibilities between the ministries separated the spinning and weaving project, and effectively the dairy project, from the more overtly agricultural aspects of the goat and sheep production system. Therefore it has to be concluded that the government, albeit inadvertently, had failed to create an economic or organisational context in which the new producers could thrive. We had thereby learned that the relationship of the Project with the government, at least as much as with the local communities, is of central importance if a rural development project's objectives are to be achieved. The theme is explored in the final chapter of this volume. In the meantime, the Khabura Project's next challenge was to integrate more closely with government from 1986 onwards and to change from a sub-regional to a national programme.

5. From a sub-regional to a national programme

5.1 Integrating with government policy

Up to the end of 1985 the Project, sponsored by PDO, had evolved its objectives, topics and approaches more or less independently of MAF and the Oman government. From 1986 it was necessary, gradually, to integrate with MAF policies and objectives.

5.1.1 Government sponsorship

While it remained based at the village of al-Khabura, the Khabura Project had only a sub-regional influence, mainly within the al-Khabura *wilaya*. This had the advantage of allowing the Project to focus its inevitably limited financial and staff resources on one set of interrelated issues in one location. Thanks to the long-term support it received from PDO the Project also had sufficient time to test and demonstrate in depth the range of approaches and techniques described in preceding chapters, and to integrate within the rural community. Up to 1982 the Project's sole governmental link had been with MAF.

From as early as 1977 MAF had agreed to sponsor and fund two aspects of Durham University's total research and development programme, in addition to those sponsored by PDO. The joint honeybee project, which supported a series of specialists based at al-Khabura and Rustaq in northern Oman and at Salalah in Dhofar, was quite separate from the livestock and irrigation interests. However, the joint spinning and weaving project was an integral part of the Khabura Project in that it sought to make fuller use of livestock by-products and promote mutual self-reliance by adapting a traditional craft industry to the exigencies of a rapidly changing world. However, its governmental linkage with MAF was set to change. In 1982, cabinet decisions designed to rationalise ministerial responsibilities decreed that spinning and weaving lay outside MAF's remit. After prolonged discussion about which Ministry would take on the sponsorship (did it fall within the remit of National Heritage and Culture or of Commerce and Industry?) H.E. Khalfan Nasser al-Wuhaybi, Minister of Social Affairs and Labour, accepted responsibility for the project. Being associated with two ministries gave the Khabura Project more impact nationally. But MoSAL's national, social affairs role meant in practice that its approach to the spinning and weaving project was to protect a dying craft rather than to create a new rural enterprise.

The year 1985 was critical for the Khabura Project. Sponsorship from PDO was coming to an end in December of that year, but would the government of Oman have sufficient interest to take the Project on? Prolonged discussions seemed to be making no progress. However, a visit to the Khabura Project by the Minister of Agriculture and Fisheries, H.E. Abdul Hafith Salim Rajab, led directly to the Project's integration into the Ministry and eventually to its transition from a sub-regional to a national set of activities.

We invited the Minister to visit us on May 15th 1985 formally to open our goat/sheep dairy unit. Somewhat to our surprise he came, driving his own car, and stayed for several hours. He enjoyed the party to celebrate the opening of the dairy, at which quantities of goat's milk yoghurt and cheesecake were eaten. He talked at length to everyone about the Project, and he queried some local farmers about our worth. Evidently their comments were positive because the next day instructions were issued in the Ministry to ensure that we were included in the budget for the 1986-90 five-year plan. Thus we gained added national recognition by achieving direct sponsorship and funding from the national government. In many ways this facilitated our work and supported our role of bridging the gap between the rural community where we worked and the new, and ever growing, network of national institutions concerned with rural change. However, MAF was more narrowly interested in agriculture than in our broader concerns with rural community development. In practice, moreover, it showed little interest in supporting rural enterprises. Indeed, enterprise was inhibited as a consequence of the provision by MAF of very heavily subsidised agricultural services; services, moreover, which were neither very well targeted nor adequately supported from the centre.

5.1.2 Links with the Oman Bank for Agriculture and Fisheries

Because we were interested in promoting small businesses, in the belief that this was a key to the sustainability of rural life, we made every effort to link as closely as possible with OBAF. They showed interest in various aspects of our work including financing a rural workshop for manufacturing fibre-reinforced irrigation channels (Ch. 4.9). Also, OBAF played a central role in the discussions with the mechanisation section of MAF and the Khabura Project which would have led to an expansion of the use of rotavators in Oman under private ownership, had not the new Minister of Agriculture and Fisheries killed the scheme on coming into office in 1986 (Ch. 4.6.2).

Also, at the time when OBAF was involved in the provision of loans for farmers wishing to adopt 'modern' irrigation systems, most particularly in relation to the GMDP, its loans manager wrote to us, via MAF, asking for costs for installing our 'low pressure drip irrigation system' and for our results[1]. We provided this information, including details of how the systems could be installed by the farmer who was continuing to show such positive initiative in expanding on his own farm the system which we had originally provided. However, they were in practice unable to recommend our locally developed system or to provide work for the local installer. Instead, they opted for

---

[1] Letter dated 30/3/1988 from the Loans Manager of the Oman Bank for Agriculture and Fisheries to the Khabura Project Manager.

imported systems which were more thoroughly tested and readily available in large quantities but which were more expensive, a drain on Omani currency and required expatriate installers. They therefore squandered an opportunity to develop a local installation industry which would also have created opportunities to train local staff in its operation and maintenance and made the staff much more available to, and accountable to, the local farming community.

However, of even greater importance, we discussed with OBAF at length the options open to them for encouraging investment in the production of goats and sheep in Oman by small and medium-sized farming enterprises. This resulted in OBAF preparing a feasibility study and proposal (OBAF, 1982) which pointed out that large numbers of animals were being imported for slaughter (over RO5mn worth in 1980) and that local sheep available in the market were 'usually of poor quality standard and very highly priced'. It stated that experts in MAF and OBAF believed that programmes could be conducted to increase local sheep and goat rearing and that these programmes would be particularly adapted to small and medium-scale farming. They proposed that the programme should be articulated around:

- a regional breeding centre producing cross-breds from local females and imported males;
- small farms rearing medium-sized flocks of selected animals.

According to the proposal, the central Batina, al-Khabura to Sohar, was identified as the most appropriate place to conduct the first sub-regional programme because the large 'modern' farm at Sohar, Oman Sun Farms, had the facilities and experience to accommodate the first breeding centre and 'The Khabura Project will provide technical, advisory and monitoring services as a logical second phase of their project, taking full advantage of the experience gained in the first phase.' (p. 2). The breeding centre capital costs (mainly for the purchase of 300 Omani ewes, 10 Chios males, 100 Omani she-goats and 4 Anglo-Nubian males), were estimated at RO50,000 plus RO20,000 running costs. This was to be financed by: MAF (RO20,000), the Special Fund for Assistance to the Private Sector (RO20,000) and an OBAF loan of RO30,000 at 2%.

The Bank believed that the initial scheme should be concentrated in one sub-region of the country first, because:

- it facilitates the provision of instruction and monitoring services from one technical centre;
- it facilitates the provision of veterinary services, including regulation of pregnancy;
- it encourages participating farmers to learn from and to support each other;
- it creates the opportunity to provide shared facilities to process by-products, including wool, hair, milk and skins;
- it justifies the creation of a sub-regional marketing organisation;

- it creates a centre of obvious and visible development where extension assistants from other sub-regions can be trained.

In the eyes of the Bank al-Khabura was the appropriate sub-region, because:

- its small-farmer community is typical of the [Batina] region as a whole;
- it is centrally located within the region;
- it is an area where the farming community and their livestock have been studied in depth [by the Khabura Project];
- it is a place where the small farmers have been exposed [by the Khabura Project] to the idea of, and have expressed interest in, rearing sheep/goats in units of medium size.

OBAF also believed that participant farmers should have the skills and interest, and the area of land, to maintain up to 100-200 head of sheep or goats, with the following resultant benefits: it would concentrate the animals on fewer, better quality farms; fewer but larger herds are more accessible to veterinary care and other aids; it encourages the farmer to realise the potential worth of by-products as sources of secondary income (e.g., wool, hair, skins, milk and dung). The original farm facilities should therefore be designed with growth in mind.

OBAF advocated the general management system being researched at al-Khabura.

5.1.3 New national roles
Although OBAF's proposed livestock scheme was not approved at the time, it exerted a longer-term influence on MAF thinking. In this context a second visit by a Minister of Agriculture and Fisheries had an even more far-reaching impact on the Khabura Project than the first. Shortly after his appointment as Minister in 1986, H.E. Shaykh Mohammed Abdulla al-Hinai made an unannounced visit to the Project's farm one Friday afternoon after social visits had taken him to Sohar for the weekend. He spoke to no one on the farm but subsequent events made it clear that in general he approved of what he had seen but believed that the Project should deploy its resources more overtly to national benefit. Amongst other priorities the Minister wanted to promote livestock research and production. In discussion with his own staff, with ourselves and with others he initiated three major livestock activities in northern Oman, each with a national dimension. They included a renewal of the Rumais Livestock Research Station (RLRS), the creation of ABARC at Wadi Quriyat, and the design and implementation of a livestock 'leader farmer' scheme which went under several names but is here called the GMDP. The medium-term consequence for the Khabura Project was its closure in mid-1989. However, we were to become centrally involved in all three of the new activities. The Project's office moved from al-Khabura to a house in the Capital Region in July of that year and, sadly, our local Omani staff lost their jobs and had to rely on other sources of family income. We had encouraged MAF to retain the facilities at al-Khabura, and

perhaps use them for other purposes. Two ideas for new usage created some interested discussion: as a centre for small-farm research into appropriate 'modern' irrigation systems, or as a veterinary facility for field training paravets and the distribution of veterinary supplies. However, in practice no funds were forthcoming for either of these options, and the centre closed completely.

Meanwhile, the Khabura Project's expatriate staff joined one or other of the new projects. We were also able to secure from ODA the secondment of three Associate Professional Officers to work with us in Oman. The first, Nick Taylor, analysed all the livestock records collected over the 12-13 years of the Khabura Project. He also devised a computer based monitoring system for the livestock on the GMDP farms. The other APOs, Mark Stephens and Czech Conroy, researched, respectively, into the factors affecting the production and productivity of irrigated forage on the GMDP farms, and the economic benefits of the GMDP to the 'leader farmers' themselves.

From 1989 to 1993/4 we had responsibility for managing RLRS and ABARC. Although we had no official responsibility for the GMDP we were involved in its design and, more importantly, we made major inputs into its monitoring and evaluation, and therefore were able to make recommendations for improving its performance - though we had no authority to implement the recommendations that we made. The work at ABARC and for the GMDP are described in some detail in later sections of this chapter because they were a direct continuation of the Khabura Project's emphasis on small ruminant production systems and irrigation.

However, the work at RLRS is here summarised more briefly because the milk and milk products emphasis switched from goats and sheep to cattle, and so is of less direct relevance to this volume. The use of A.I. gradually to introduce exotic dairy blood into the local cattle population had already been proposed (El-Dessouky, 1986; Mohammadein, 1989) and previous work at RLRS had shown that the cross-bred Jersey x Billadi (local) cattle produced acceptable, and probably sustainable, milk yields well above those of the pure Billadi cows. As the first stage of a pilot project, Kate Rogerson co-ordinated a survey of some 437 cow keepers within a 30km radius of RLRS. The survey focused on the number and types of cows owned, their milk production and usage, and the ways in which the cows were mated. The concept of A.I. was also explained to the farmers, and they were asked whether they would use it if it were available. A sub-text to this last question was to find out if a significant number of farmers had any scruples about using A.I. on religious grounds.

The survey results (Rogerson and Omer Binofuf, 1990) showed that most of the cows were Billadi and that most farmers owned one or two head only. Over 90% of the owners wanted to produce more milk. Almost everyone consumed, at home, all the milk they produced though a handful of more commercially orientated herds also sold fresh milk and many people sold *semmen*[2]. Two-

---

[2] *Semmen* is defined by Rogerson and Omer Binofuf (1990) as ripened and spiced ghee.

thirds of the cows were served for a fee, typically of RO4-5, by bulls owned by other people, showing that owners were prepared to pay for breeding services. No one had experience of A.I. but three-quarters of the farmers were interested in the idea particularly if it would improve productivity. Also, some people reported, service bulls were less easy to find nowadays. Only 10 people expressed a religious reason for lack of willingness to participate in the proposed service.

Because of the high degree of interest expressed by the cow owners, the A.I. scheme was introduced at the end of 1990, and by the end of 1993 some 1,751 inseminations had been carried out. Ward (1994a) concluded that the pilot project had been successfully initiated in terms of farmer interest (430 participants by the end of 1993 and new ones still registering at the rate of eight per month), number of inseminations (1,751), pregnancy rates (exceeding similar schemes by 10%) and calves born (422). But the full impact of the scheme had to await the results of completed lactations. A key to success was the daily 'presence' of the scheme in all the villages in which the farmers lived. An inseminator followed the same route each day to collect the tokens left by the farmers in the A.I. 'post-boxes' in each village. Thus the service was reliably delivered at the appropriate time in the cow's cycle. Three months later, pregnancy diagnosis was undertaken, and the inseminator made another visit after the birth of the calf. Daily action lists were produced by the computer database, and a copy of the record was left with the farmer. Other reasons for success were the use of top quality proven semen sources and well-trained staff available seven days a week. Plans to expand the scheme were under discussion in 1994, starting in Nizwa.

Although this work had no direct immediate planned reference to goats and sheep, if, at a later date, there is a desire to explore nationally the potential of goats (and perhaps sheep) for milk production the positive results of the early years of running the goat milk scheme as part of the Khabura Project, coupled with the growing experience of the cow A.I./dairy project, will form a good basis for structuring a milk scheme based on village or sub-regional dairies.

5.2 Wadi Quriyat Animal Breeding and Applied Research Centre

Although the Minister was not convinced of the value of goat dairy schemes he was nevertheless determined greatly to increase meat production from goats and sheep in northern Oman. He wanted to ensure that goats and sheep were efficiently managed on small and large production farms to fulfil Oman's growing meat demand and therefore reduce the need for importation. The Minister was also concerned that the sheep and goats, particularly the goats, owned by the *shawawi* pastoralists were 'roaming freely' in the mountains and on the plains of the Batina and interior Oman where they were eating and putting at risk the sparse vegetation whilst yielding few progeny and little meat. It was argued that if these animals could be fed more effectively from other sources they would both conserve Oman's wasting reserves of natural vegetation, and become more productive.

It was already envisaged that the Ministry farm at Wadi Quriyat would play a major role in expanding the number of sheep and goats, and that there would be other multiplication centres, large and small, throughout northern Oman. In response to a request from the Ministry to assist with the then proposed 'thousand goat distribution project' for northern Oman, the Khabura Project submitted a study including livestock housing designs together with details of systems for forage production and irrigation.

Production records at the Khabura Project and other research and monitoring centres in Oman had shown that local breeds of goats and sheep exhibited some good production traits but with a high degree of variability. The animals were also valued above imported stock by the people, and they were adapted to Omani conditions. This combination of characteristics made them suitable for a selection and breeding programme to create recognised improved breeds for rearing by farmers. It was therefore decided by the Ministry to establish a new unit at Wadi Quriyat (ABARC) containing sizeable flocks of goats and sheep, representing the main types found in Oman, with the following objectives (Ward, 1994b):

- to obtain basic information on the characteristics and productivity of the main local types of goats and sheep;
- to implement a scientific breeding and selection programme for the improvement of the local goat and sheep types, so producing recognised breeds;
- to produce and distribute improved goat and sheep breeding stock to farmers all over the country, particularly to those farmers participating in government sponsored livestock projects.

During 1987/8 Khabura Project staff had the opportunity to comment on and suggest improvements to the architectural plans for ABARC's new buildings, and stressed the requirement for good veterinary health facilities. The buildings were mainly constructed during 1989, and the foundation stock purchased during 1990.

5.2.1 The goat and sheep breeding system
The first phase (1990-3) of the research focused on breed evaluation through performance testing of the breeding animals and their progeny. The most easily identifiable goat types were the long-haired, variously coloured goats found on the Batina and in other parts of lowland northern Oman (and referred to by ABARC as Batina goats), the somewhat larger, tawny-coloured goat found in the Jabal Akhdhar, and the small, mainly white, short-haired goats characteristic of the southern province of Dhofar. It was not possible to distinguish different types of sheep. During 1990 up to 300 of each goat type were purchased including a small number of males, also 292 sheep including 12 males.

184

The main, government-sponsored livestock project designed to improve small ruminant management and productivity on small commercial farms was the GMDP, and the proposed equivalent sheep project. ABARC would provide genetically improved stock to the GMDP farmers (amongst others). Therefore the link between ABARC and the GMDP was overt from the outset.

A rigorously designed breeding programme was followed (al-Nakib, 1994) to test the performance of dams and progeny. Key economic indicators used in the testing were number of progeny born, and growth rate to weaning which, in the breeding programme, was at three months. These indicators reflected the doe's prolificacy and mothering ability. A third criterion was the kid/lamb weight at six months which reflected its inherent ability. The doe's adult body weight was also included in the selection index. According to the responses predicted for the selection index, after 10 years the level of twinning should reach 75% , and the kids and lambs should be approximately 2.5kg heavier at weaning at the same age as before. However, it was also recognised that at the expected mortality rate of 15% the rate of increase in the number of progeny weaned would be 50%, an acceptable rate of improvement. If we assume that progeny were weaned at 10kg at the start of the breeding programme and that this rises to 12.5kg after 10 years the annual gain in meat production will amount to 8.8 tonnes. This increase will derive from the flock of 1,200 breeding females together with the production of the estimated 460 breeding males and females distributed annually to participating farmers, and the sale of the 300 breeding males and females culled annually.

At the end of the 6-month progeny testing period, the top performing males and females were added to the nuclear flock (replacing poor performing or older animals which were culled). The next ranking males and females were distributed to participating farmers and the remaining animals of both sexes were sold for slaughter. In this way the most productive genes were retained in the flock while the poorest genes were eliminated from the system. One of the most interesting results from the first three years (al-Nakib, 1994) was the evidence for the superiority of sheep over goats in output (26-33%) and efficiency of production in terms of output per unit of metabolic bodyweight (35%). The sheep superiority derived from their higher fertility (85% v. 81%, a 5% advantage) and lamb weaning weight (16.2kg v. 12.2kg, a 33% advantage)[3], which more than compensated for the inferiority in the ewe's litter size weaned (0.97 v. 1.02, or 5%). As the sheep and goats had all been randomly purchased from similar environments in different parts of Oman within the same time period when they were of poor to moderate body condition, and as they had then all been treated identically at ABARC, al-Nakib (1994) was able to conclude that sheep are more responsive than goats to a higher level of nutrition and management. He recommended therefore that sheep should in the future be

---

[3] This reflected results on the Khabura Project where at three months, local lambs weighed 15.0kg, and kids 12.3kg (combined male and female means) (Taylor, 1991a).

taken more seriously than in the past, which reflected the findings and conclusions from the Khabura Project[4]. Sheep are also easier to manage in the confined spaces of pens. Although there is a traditional bias against sheep in Oman it is paradoxical that most small ruminants imported for meat are sheep. Thus tastes are changing, and the opportunity to exploit the sheep's productive potential should be explored.

Amongst the goats at ABARC, the Dhofari goat was superior to the others in consequence of its higher litter size weaned and its smaller body weight. The Batina goat (the type reared on the Khabura Project's farm) was a more efficient producer than the Jebel Akhdhar type, and thus the latter, from the evidence at ABARC, was the least interesting animal of them all. Yet this type is superficially perhaps the most attractive - tall and tawny coloured - which is why it was the preferred goat for the GMDP scheme.

Table 29. Annual breeding and stock management cycle

Month	Activity
September/October	Foot trimming pre-breeding
	Selection of replacements, culls and animals for distribution
	Flushing of breeding females
October/ November	Breeding
February	Preparturition: increase feeding of breeding females
March/ April	Kidding/lambing
May/June	Clipping/shearing
	Weaning
	Foot trimming
June September	Progeny test

Source: Kwantes, 1994

The breeding and stock management cycle followed at ABARC is outlined in Table 29. An annual breeding cycle was adopted in order to minimise stress on the adult stock. Breeding was natural, with the males put to the females in October/November, so that all kidding and lambing occurred in March/April. This facilitated comparison between types and individuals because the environmental effects on all animals were similar at each stage of pregnancy and kid/lamb growth. Work at the Khabura Project and elsewhere had shown the winter months to be a time when fertility was high and environmental stress on the animals at its lowest (Taylor, 1990a).

The project was on a grand scale, as is indicated in Table 30 by the total number of adult goats and sheep on the station. The overall head of stock, including followers and young stock awaiting distribution to the participant farmers, had risen to 2,500 in July 1993.

---

[4] From the Khabura Project results Taylor (1991a) concluded that local sheep were superior to local goats in producing offspring body weight because: (a) lambs gained weight faster in the first year, (b) breeding ewes have a smaller mature size, (c) lower age at first conception, (d) a shorter breeding interval, and (e) slightly better survival in housed conditions.

Table 30. Adult goats and sheep at ABARC - summary statistics, 1990/1-1992/3

Yr		Batina goats		Dhofar goats		Jebel Akhdhar goats		Omani sheep		Total
		F	M	F	M	F	M	F	M	
0	30.9.90	222	25	279	23	308	20	282	16	1,175
1	Deaths	3	1	9	2	8	0	3	0	26
1	Distribution	0	0	0	1	0	0	0	0	1
1	Sold	1	2	6	0	4	1	5	0	19
1	30.9.91	218	22	264	20	296	19	274	16	1,129
2	Deaths	10	0	15	1	9	1	4	0	40
2	Distribution	0	11	0	8	0	9	0	11	39
2	Sold	14	3	34	5	15	6	20	3	100
2	Replacements	43	22	58	21	46	25	53	20	288
3	30.9.92	237	30	273	27	318	28	303	22	1,238
3	Deaths	8	1	10	1	12	2	5	0	39
3	Distribution	3	0	23	0	22	0	23	0	71
3	For sale	40	0	47	0	74	0	38	0	199
3	Replacements	76	20	78	16	87	2	78	19	376
3	Jul 1993	262	49	271	42	297	28	315	41	1,305

Source: Data from Kwantes, 1994

5.2.2 Management and veterinary care

The general management logistics, successively the responsibility of Robin Jackson and George Sidgwick, were severe (Sidgwick, 1994), particularly during kidding and lambing in March/April each year. The management lessons learned were of great relevance to large commercial farms. But they were not of much relevance to the participant farmers on the GMDP scheme because of the huge difference of scale of operation.

The record of veterinary conditions is, however, of great relevance to the GMDP farms. Although, in part, the occurrence of diseases and mortality were associated with the unique concentration of so many small ruminants in a confined space, the problems found were common to all regions of Oman. Their occurrence at ABARC and their methods of control are a very important source of information for the veterinarians in other parts of the country, including the veterinary administration in MAF. They have the responsibility for disease control on the GMDP and others farms to which the improved young stock from ABARC are being supplied.

The major diseases noted in adults in each year, 1990/1-1992/3, are listed in Table 31. Diarrhoea and abscesses were major problems in all three years. Conjunctivitis was brought under control after the first year, but mastitis and injuries became more common. There were also some important breed differences, notably the susceptibility of Dhofari and Jebel Akhdhar goats to abscesses and the relative frequency of mastitis in sheep. The group 'other' forms a high proportion of the total, indicating that the range of problems was very great. Pneumonias and respiratory tract diseases did not figure in the list at all, in contrast with both the Khabura Project veterinary clinic record (Dutton,

1994), the Khabura Project's farm (Taylor, 1991a) and the GMDP farms (Kwantes, 1993) where it was a common. At ABARC a combination of vaccination and isolation protected the animals very effectively. Equally interestingly, the sheep suffered many fewer disease problems than the goats. Finally, the number of adult stock that died in each year, both sheep and goats, was low.

Table 31. Disease presence at ABARC in adult goats (and sheep) 1990/1-1992/3[1]

		Goats									
		Batina		Dhofari		J. Akhdhar		Omani Sheep		Total	
Years	Diagnosis	n	%	n	%	n	%	n	%	n	%
1-3	Abscess	28	7.7	180	27.5	162	27.3	16	6.6	386	20.8
1-3	Diarrhoea	68	18.7	55	8.4	75	12.6	6	2.5	204	11.0
2 & 3	Mastitis	35	9.6	54	8.3	47	7.9	44	18.2	180	9.7
2 & 3	Injury	28	7.7	37	5.7	57	9.6	23	9.5	145	7.8
1 & 3	Abortion	17	4.7	28	4.3	26	4.4	25	10.3	96	5.2
1	Conjuncti'is	34	9.3	19	2.9	27	4.5	1	0.4	81	4.4
3	Dystocia	5	1.4	12	1.8	5	0.8	3	1.2	25	1.3
1-3	Other	149	40.9	269	41.1	195	32.8	124	51.2	737	39.8
1-3	Total	364	100.0	654	100.0	594	100.0	242	100.0	1,854	100.0

Source: Data from Kwantes, 1994

(1) From tables listing the 4 most common diseases in each year

Table 32. Disease presence at ABARC in kids and lambs, 1990/1-1992/3[1]

		Goats									
		Batina		Dhofari		J. Akhdhar		Omani Sheep		Total	
Years	Diagnosis	n	%	n	%	n	%	n	%	n	%
1-3	Diarrhoea/										
	Enteritis	176	60.7	101	40.1	234	48.3	45	18.4	556	43.7
1-3	Ill-thrift	61	21.0	62	24.6	143	29.5	143	58.4	409	32.2
2 & 3	Injury	9	3.1	17	6.7	36	7.4	10	4.1	72	5.7
2 & 3	Arthritis	5	1.7	21	8.3	6	1.2	0	0.0	32	2.5
1	Pneumonia	2	0.7	1	0.4	6	1.2	0	0.0	9	0.7
1	Septicaemia	2	0.7	4	1.6	3	0.6	0	0.0	9	0.7
1-3	Other	35	12.1	44	17.5	56	11.6	22	9.0	157	12.4
	Total	290	100.0	250	100.0	484	100.0	220	100.0	1,244	100.0

Source: Data from Kwantes, 1994

(1) From tables listing the 4 most common diseases in each year

Table 32 lists the major disease problems affecting kids and lambs in each year showing that the dominant problems in each were diarrhoea/enteritis and ill-thrift. Apart from these two, injuries and arthritis caused some problems in years two and three. As in the case of the adults, pneumonia was relatively unimportant. The group 'other' is quite a small proportion of the total, indicating that the range of kid and lamb ailments was low.

The proportion of kids and lambs suffering from one or more ailments is noted in Table 33. The proportion is high, and climbed rapidly after year one, with a mean of 42%. The high mean and the increase after the first year will be associated with the confined conditions in which the stock were housed. The sheep once again show their superiority, their lambs suffered only half the problems of the Jebel Akhdhar kids, and less than the other types of kids as well.

Coincidentally (or perhaps not coincidentally) the order of resistance to disease is the same as the overall order of performance superiority (al-Nakib, 1994), as stated above.

Table 33. Proportion of ABARC kids and lambs with disease problems, 1990/1-1992/3

Year	Batina goats		Dhofar goats		J. Akh. goats		Sheep		Total	
	kids	% ill	kids	% ill	kids	% ill	lambs	% ill	all	% ill
90/1	137	21	190	17	196	26	204	32	727	25
91/2	227	54	234	45	309	71	295	22	1,065	48
92/3	234	59	269	42	305	70	325	35	1,133	51
Total	598	48	693	34	810	59	824	30	2,925	42

Source: Data from Kwantes, 1994

Table 34. Causes of death in kids/lambs at ABARC, 1990/1-1992/3

Years	Cause of death	Batina goats		Dhofar goats		J. Akh. goats		Sheep		Total	
		n	%	n	%	n	%	n	%	ttl	%
1-3	Ill-thrift total	68	32.4	64	29.4	137	41.9	150	72.8	419	43.6
1-3	Diarrhoea/Enteritis	58	27.6	45	20.6	88	26.9	14	6.8	205	21.3
1	Conjunctivitis	27	12.9	8	3.7	15	4.6	6	2.9	56	5.8
1	Arthritis/Septicae.	4	1.9	35	16.1	16	4.9	0	0.0	55	5.7
2	Pneumonia	0	0.0	0	0.0	6	1.8	0	0.0	6	0.6
2	Intestinal/Accident	1	0.5	1	0.5	3	0.9	0	0.0	5	0.5
3	Injury	1	0.5	2	0.9	1	0.3	1	0.5	5	0.5
3	Peritonitis	1	0.5	1	0.5	0	0.0	1	0.5	3	0.3
3	Septicaemia	0	0.0	2	0.9	0	0.0	1	0.5	3	0.3
1-3	Other	50	23.8	60	27.5	61	18.7	33	16.0	204	21.1
	Total deaths	210	100.0	218	100.0	327	100.0	206	100.0	961	100.0

Source: Data from Kwantes, 1994

Not surprisingly, given the occurrence of disease in kids and lambs noted in Table 34, the main causes of their deaths in each year were ill-thrift and diarrhoea/enteritis, though enteritis was effectively brought under control after the first year. In year one, conjunctivitis and arthritis were also important causes of death, but were then brought under control. The group 'other' accounted for one-fifth of all lamb and kid deaths. Importantly, it seems that lambs are vulnerable to the rather ill-defined and understood disease, ill-thrift, but if they survive it they are resistant to most other ailments. Kwantes also described the symptoms he witnessed as 'poor do-er', 'lethargy', 'emaciation' and 'non-specific deaths'. The records at the Khabura Project's farm also revealed that a high proportion of sheep deaths were similarly categorised as 'still births' and 'weak lambs'. Taylor (1990b) noted that this could have been nutrition/management related (but why then was it much less significant in the goats?) or: 'there could have been some underlying disease component (eg toxoplasmosis, which can typically cause a complex of late abortion, stillbirth and weak lambs)' (p. 52).

The rate of kid and lamb mortalities (Table 35) remained stubbornly high, largely due to ill-thrift, and was higher than the mortality assumption of 15%

made by al-Nakib and upon which he estimated a 50% increase in progeny weaned after 10 years.

Table 35. Kid and lamb mortalities up to 6 months at ABARC, 1990/1-1992/3

Years	kid/lamb deaths	Batina goats	Dhofar goats	J. Akh. goats	Sheep
1-3	Total kids/lambs	607	741	825	830
1-3	Total deaths	93	94	162	154
1-3	% deaths	15	13	20	19
1	% deaths	21	16	26	20
2	% deaths	15	13	17	16
3	% deaths	13	12	20	20

Source: Data from Kwantes, 1994

5.3 The Goat Multiplication and Development Project

The Khabura Project's farm and the livestock schemes of the participant farmers in the region of al-Khabura who had adopted a simplified version of the Project's production system, were the basis of the GMDP design. For example, the Project provided basic architectural drawings for simple, low-cost stock housing, and the MAF architect was sent to Khabura to make detailed measurements of livestock pen dimensions and infrastructure. Designs to house up to 40 breeding female goats, plus a few males and followers were then drawn up for the GMDP farmers. The first trial units were built at al-Hamra. Subsequent units incorporated important design modifications suggested by the Khabura Project staff to simplify and improve their ease of operation and usage.

The sprinkler systems to irrigate the forage crops were approved during 1987 by MAF and OBAF. The Khabura Project engineer, Nick Foster, participated in the evaluation discussions. He also undertook technical surveys of selected farms and agreed to train staff from the extension centres so that they would be able to participate in the remaining technical surveys. It is largely as a consequence of MAF's decision to use sprinkler irrigation systems for the forage crops on the GMDP farms that the Khabura Project decided to introduce sprinkler irrigation to part of its own forage area in 1988. It was hoped that its effect on crop yield would be long-term monitored, and the system improved as required, and demonstrated to local farmers and MAF extension staff on training courses to be arranged during 1988 and beyond. In practice, however, such training was limited because of the decision to close the Khabura Project by mid-1989 in order for the staff to be more actively involved in ABARC, RLRS and the GMDP (Ch. 5.1.3). Unfortunately, an important appropriate trial, monitoring, evaluation and training facility was thereby lost.

Many views were incorporated into the final GMDP design, including those of the Minister who decided that only goats should be involved (no sheep), and that the Rhodes grass should be sprinkler irrigated. The Minister also insisted that the goat pens should be built from welded steel piping roofed with corrugated asbestos sheeting, instead of the timber which we had bought in the *suq* roofed with locally made *barusti* sheeting.

The GMDP's terms of reference were compiled in 1987 by the Director of the MAF's Department of Animal Wealth (Khamfar, 1987). At this stage it was envisaged that sheep also would be in the scheme. The aim was to: 'motivate the breeders and farmers to contribute and participate in animal production projects in order to increase the national production of goats and sheep and upgrade the farmers' efficiency by enabling them to maintain self-reliance in utilising and developing the natural resources in the Sultanate through improved methods of production and breeding'. The project's aims were:

1. Establish sample production units for goat breeding with the purpose of developing and increasing the quantity of local goats and their products of meat and milk throughout the northern part of the Sultanate.
2. Subsidise and encourage the farmers to undertake investment in the field of animal production, particularly goat and sheep breeding.
3. Develop local resources and increase the local production of goats and sheep, to achieve a reduction in the quantity of imported meat (80% of total consumption) and save hard currency.
4. Breed goats and sheep using appropriate scientific and technical methods under the continuous supervision and follow up of those who are concerned in MAF.
5. Reduce the animals' market price due to the increase of supply.
6. Increase per capita income due to the revenue generated from goat and sheep breeding.
7. Build enclosures which are an ideal example for those who are interested in goat breeding.

The GMDP scheme was originally devised for 30 farmers in the al-Khabura region, then expanded to 150 farmers selected from all parts of northern Oman according to their interest in and experience with small ruminants. Finally, because 1988 was declared Agriculture Year, the number of farmers to be included in the first phase of the scheme was doubled to 300, and later almost redoubled to 550 to be equipped by 1992/3. They were to be provided free of charge with 10 good-quality local female goats four months pregnant by local males, and one mature male. So, the Ministry had first the task of buying 3,000 female and 300 male goats for subsequent health checks, mating and distribution. The 'Durham University Research Team at al-Khabura' was given an important follow-up role, but it had no role in the decision to introduce the scheme (except by example at al-Khabura) nor in the selection of the participant farmers. In practice the selected farmers tended to be the more influential and affluent ones - partly because only they could be sure of being able to repay the irrigation system bank loans, see below, because they had other sources of income if the project failed. In the Wadi Quriyat area reluctant farmers had to be personally encouraged by the Minister to join - he was enthusiastic because his home village is in the region. Their reluctance was in no small measure due to the fact that although they were getting goats and pens (designed and erected by MAF) free of charge, they had to agree to install pumpsets and piping suitable to operate sprinklers to irrigate two feddan (0.8ha) of Rhodes grass. OBAF loaned RO3,000-4,000 to the farmers for the irrigation systems, but the loans had to be repaid over about six years. Originally the only forage was to be Rhodes grass but at the Minster's personal insistence an area under the sprinklers was set aside for barley to be grown for grain as a concentrate feed.

There were various weaknesses in the system: in practice many of the Omani farmers (who were supposed to be hands on) were absentee or semi-absentee, and much of the work was undertaken by labourers from the Indian sub-continent with no knowledge of or interest in the scheme or livestock, and equally little knowledge of Arabic; there were small but significant faults in pen design; the quality of the animals provided to the farmers was variable; from the outset it was difficult to see how the farmers could repay their OBAF loans; the extension and veterinary services probably would not have been adequate to cope with severe problems should these have occurred; many of the pumps had insufficient power to operate sprinklers effectively; and no one, other than the Minster, had faith in the idea of growing barley for grain as part of the project.

However, it was a significant achievement that the introduction of the scheme went more or less according to timetable; by the end of winter 1988/9 practically all the pens had been completed. Most farmers had received their allocation of goats, some of which had already kidded in their new pens. Also, the presence of the scheme provided the Ministry with a sense of direction about improving its veterinary and livestock extension services. The goats, in general, appeared to be thriving. In spite of the weaknesses in the design of the irrigation system, the forage crops were in many cases growing reasonably well considering that Rhodes grass was new to many of the farmers and using sprinklers new to them all. A few farmers appeared to be doing very well, others not so well and others badly. The scheme provided everyone involved in it with a wide range of interlinked experience from which many valuable lessons could be learned, and MAF, in principle, was aware of this and encouraged us to organise a monitoring and evaluation system in order that strong points could be more widely known, and weaknesses overcome. Up to the time that the Khabura Project's farm closed in mid-1989 we kept a fairly close eye on the 12 GMDP farmers near al-Khabura, particularly on their forage crops and irrigation systems. Based on this experience we believed that most problems affecting the forages were due to fairly obvious things like poor distribution of water, shortage of nitrogen and unfamiliarity with the system.

5.3.1 *Managing sprinkler irrigation*
Having decided that Rhodes grass would be the principal forage crop, the Minister also decided that it would be sprinkler irrigated. After he took office in

1986 the Ministry gave increasing thought to improving irrigation efficiency. The general conclusion, not necessarily valid, was to encourage farmers to change from flood irrigation to the various systems by which water could be piped directly to the plants. Under these circumstances it was almost inevitable that for field crops, such as Rhodes grass, on small farms, hand moved sprinkler systems would be recommended. Thus it was decided to install them on all 300 of the 'leader farmer' GMDP farms when they were being equipped during 1988 and thereafter. This entailed extensive survey work, in which the Khabura Project staff played an important role by pointing out that the typical pumps used to lift water for flood irrigation were not powerful enough to generate the pressure required to operate sprinklers efficiently.

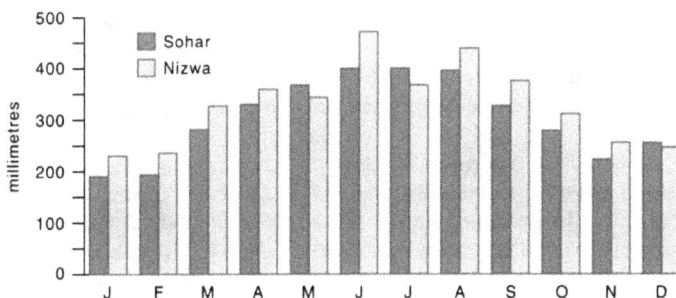

Source: Foster, 1989

Figure 23. Monthly irrigation rates, Sohar and Nizwa

We were also asked by MAF to produce an irrigation schedule for the GMDP scheme. The work was undertaken by Foster (1989) using a modified Blaney-Criddle formula, and based on climatic data from five Agromet stations that had been installed in the different regions of northern Oman, and whatever soil data that could be found. Figure 23 illustrates the monthly totals recommended for two towns in contrasting climatic zones, Sohar on the Batina coast and Nizwa in the northern interior. They are cautious in that they are based on the highest recorded ETo for each month and include a leaching factor. The annual recommended totals for Nizwa and Sohar were 3,972mm and 3,658mm respectively - almost 4m depth but considerably less than the depths applied by typical farms using flood irrigation (Dutton, 1991).

It was recognised by Foster that the proper operation of the sprinkler system, in order to achieve the desired increases in water use efficiency, implied a level of technical sophistication and interest in water conservation issues that was completely beyond the experience of almost all the farmers in Oman, and perhaps of all the farmers in the scheme. To make it slightly more simple he devised an irrigation schedule using a set three day interval, so that the farmers would only have to vary the sprinkler operating times at different seasons of the year.

Foster and others (Dutton and Abdul Baqi, 1992) also compiled a very informative illustrated irrigation manual with sections on basic rates and methods of measurement, on a simple theory of irrigation engineering, on sprinkler irrigation specifically relevant to the GMDP, and on forage husbandry. We also translated the whole manual into Arabic and had it properly printed and bound to encourage its usage by MAF extension staff. It was well received but not used to anything like its potential as a training tool because it was never actively advocated by the central MAF officers. Also, although it was made as simple as possible it was still well beyond the level of the farmers, who were mostly illiterate, and even further out of the reach of the field operatives on the farms who were typically from Bangladesh and had little or no knowledge of English or Arabic, or indeed irrigation. Although we suggested that the manual should be translated into Bengali, MAF found this to be politically unacceptable, and in any case most of the Bangladeshis were not literate in their own language either.

In January 1993 Stephens (1993a) was helped by the Regional Animal Production Officers[5] in the extension centres to obtain answers to four very basic questions from each of the 300 farms in phase 1 of the GMDP: is Rhodes grass grown; is the irrigation system flood or sprinkler; is the sprinkler system now in use; and if not in use, why not. Growing Rhodes grass under sprinkler irrigation was, of course, central to the project concept. However, 42% of the farmers were not growing any Rhodes grass, and as many as 54% were not using the sprinkler system supplied by the GMDP. Of those not using the sprinkler system, just over half gave lack of water as the reason. No one on the Batina gave this reason, but by contrast a remarkable 53-83% of farmers in different parts of the interior stated that there was not enough water for the sprinkler system to operate. It would seem that a more careful approach to farm selection on the basis of well yields, recharge rates and existing farm requirements would have shown the majority of those farms to have been unsuitable at the outset of the project, and thereby saved a great deal of money, time and effort. As it was, confidence in MAF was undermined, and the 'leader farmer' principle placed in jeopardy. On the Batina coast, where no one complained of water shortage, but over 40% of the farmers were not using sprinklers to grow Rhodes grass the main reasons given were leaking pipes or broken pumpsets. One cause of leaking pipes was low water operating pressure; the pipe sealing rings are only effective if the pressure is sufficiently high. Other pipes were holed or otherwise damaged and not repaired. But whatever the cause of the machine and pipe failures it is clear that the Batina farmers were not sufficiently interested in the sprinkler system, and the use of it to grow Rhodes grass, to maintain it.

---

[5] Appointed specially to provide support to the GMDP leader farmers.

The correlation between the findings of the questionnaire survey of January 1993 and the survey of 99 of the 300 farms a year earlier (Stephens, 1992) was highly significant both for the types of irrigation used and the areas of Rhodes grass grown on the different farms. From the latter it may also be deduced that in the year between the surveys there was no significant increase in Rhodes grass area. The project intention was that two feddan or 0.8ha would be sown with Rhodes grass but in fact the stable mean area was only half (0.39ha) on those farms actually growing the grass. Overall, the mean proportion cropped with Rhodes was only 22% of the design area.

The low proportion of land under the crop was partly related to the problems of water shortage and mechanical/pipe failure mentioned above, but it was also connected with the way in which, in practice, the farmers used their sprinkler layout. They mostly opted for the cheaper installation of two manually moveable laterals each with 8 sprinklers that covered one-third of the planned two feddan. If this was moved each day, and Foster's 3-day irrigation schedule followed, then two feddan of Rhodes could have been grown. However, only two surveyed farms were seen using the system in this way. The remainder used it as a fixed system, which of course restricted to one-third the area they could sow and irrigate.

Other reasons for the small area of Rhodes grass under sprinklers seemed to be associated with the number of goats and the traditional preference for feeding the goats with alfalfa. The mean number of adult goats in the intensively monitored goat units was less than half of the design capacity, according to Conroy (1992c). Conroy also noted that only one-quarter of the phase 1 GMDP farmers fed Rhodes as the main component of the ration; in general, alfalfa was preferred. The suitability of using Rhodes grass as a feed, and its production advantages over alfalfa, were neither explained nor demonstrated to the farmers which again was undermining the central concept of establishing a 'leader farmer' scheme.

One objective of Stephens' 1993 survey of all 300 farms was to test the validity of his findings of a year previously, when he personally visited 99 of the farms. We have already seen that in terms of irrigation system being used and area of Rhodes grass grown, little had changed during the year, and that the 99 farms were fully representative of the 300 total in the scheme. This makes the more detailed findings of the 99 farm survey all the more significant. Stephens (1992) noted that only two of the pumps seen were known to have been bought specifically with sprinklers in mind. Of the others he notes that all pumps were adequate for pumping to cisterns, though the condition of many would have made them very inefficient even for this undemanding low-pressure task. But sprinklers require much higher pump pressures in order to operate effectively, and the design pressure of 2.76 bar at sprinkler would not have been attainable by many of the pumpsets even if the farmers had understood the importance of doing so. Indeed Stephens stated: 'Virtually all sprinkler systems showed evidence of being operated well below the design pressure' and 'motors powering the systems were often being run at very low rpm and consequently there is no hope of attaining the design pressure' (p. 7). Stephens states that with a design pressure of 2.76 bar, an operating pressure above 1.5 bar would have given an adequate distribution pattern. However, on the 12-farm detailed study (Ch. 5.3.2), only four farms were operating above this modest level, and five well below it.

In fact there were very few systems on the 99 farms where the pressure gauge worked or was readable, or the reading bore much relationship to the outlet pressure. Stephens (1993b) reported that: 'There seemed to be an almost total lack of awareness amongst the farm staff who operate the system, that the operating pressure was of major importance in the efficient management of the system' (p. 42), and he concluded: 'using a modern irrigation system in such a way makes a mockery of the drive towards more efficient irrigation on small farms' (p. 42).

When sprinklers are operated well below their design pressure it appears (to the uninformed) that the problem of water distribution lies with the nozzles. This explains the farmer response. In 31 cases the secondary 'slit' nozzle, which normally restricts water to the inner part of the irrigation circle, had been replaced with another round nozzle which normally throws water to the outer part, and in five cases most or all the nozzles had been completely removed. Given low pump pressure, these measures may have maximised total water delivery but with no possibility of achieving an even distribution.

Not surprisingly, the farmers, by their actions, did not show any enthusiasm for sprinkler irrigation. In fact, while 51 of the 99 farmers were growing some Rhodes under sprinklers 10 were using flood irrigation and, in the case of alfalfa, 83 were growing it under the traditional flood system and only seven were test growing some under sprinklers. Also, 40 farms had never used or had stopped using the sprinkler system altogether.

On most farms the sprinkler system extended the irrigated forage area, instead of replacing existing flood irrigated forage crops. Therefore, instead of reducing the total water demand, the sprinkler system increased it thereby putting additional strain on the scarce water resources which had prompted the initial investment in the scheme. Worse, it was known that many of the farms selected for the scheme, especially in the northern interior of the country, had a history of inadequate water supply. In these cases it was therefore particularly unfortunate that the scheme was resulting in attempts to increase forage area and water consumption, not reduce them. A more rigorous selection of the farms entering the scheme, as stated above, should have eliminated those without adequate water before any capital investment had been made.

However, it is important to note that examples of good management were found including farm DD2, at Dhank, in northern interior Oman, where irrigation application rates followed seasonal variations in calculated water requirements. The owner visited the farm most days and employed an Egyptian foreman/worker who had a good knowledge of both crop and animal husbandry. Rhodes was fed as the main fodder crop. Unfortunately, even their best efforts were undermined in March 1993 when well recharge rates dramatically declined following a large wadi flow.

5.3.2 Rhodes grass yields
The management of the Rhodes grass on 12 of the GMDP farms was the subject of a detailed year-long study by Stephens (1993b). He established sets of three evaluation plots within each farmer's field. Each plot contained a home-made 'rain gauge' constructed from a mineral water bottle, a length of hose pipe and a four litre oil can. The rain gauge was set between two $1m^2$ patches of Rhodes, one given extra fertiliser and the other the control. One plot was placed equidistant from four sprinklers, a second equidistant from two sprinklers and the third 6m from one sprinkler. The patches on each plot were cut at approximately the same time that the farmer cut the rest of his field. Thus it was possible to estimate yield under two different fertiliser treatments, and three different irrigation regimes. It was also possible to estimate the total irrigation water applied and its distribution in relation to the proximity of one or more sprinkler heads; in theory the water distribution should have been equal across the field. Soil and water samples were also analysed.

When Stephens (1992) made his 99 farm survey he also judged the level of management of the Rhodes grass by visual inspection of the sward density, uniformity and colour. Using these criteria he produced the following classification, and numbers of farms in each class:

Very good	8
Good	13
Adequate	17
Poor	13
Dead or burnt	13
Not grown	35

Perhaps the most interesting point, if one wishes to be optimistic, is that 21% of the Rhodes fields were good or better, including 8% which were very good. It was therefore possible for small farmers, given adequate natural resources, to produce very good forage crops. If the reasons for their success could be identified, perhaps other farmers could be encouraged to emulate them. However, it must not be forgotten that at the other end of the scale, 13% of the fields were poor, and 48% of the farmers had either given up or had never started.

Originally 17 farmers were selected for the year-long detailed study, and it is perhaps interesting to record the reasons why five of them were dropped because it illustrates the diversity of problems not only affecting the study but also affecting the Rhodes production scheme. The reasons were, respectively:

197

animals were randomly grazing the field; severe water problems; the Rhodes crop was abandoned with a view to resowing at a later date; extremely saline water had killed the Rhodes grass; a broken pump was replaced by one which was not powerful enough to drive the sprinklers at all.

The 12 farmers who remained in the study were all growers, but otherwise reflected the range of success witnessed in the 99-farm study. Five farms in the evaluation exceeded 20t DM[6]/ha, which is very acceptable, but the remainder achieved less than 20t DM/ha, including one less than 5t DM/ha. On the patches to which an additional 120kg N/ha/cut were added, four yields were in excess of 30t DM/ha, the highest reaching the remarkable 53 tonnes. But what are the reasons for these yield differences?

A general cause of low yields was the poor nutrient status of the soil. This was demonstrated in the case of nitrogen by comparing the yields from the farmer-fertilised patches in the evaluation plots with those from the patches to which 120kg N/ha/cut was added. The yields in the latter were 24% higher than in the former. Additional nutrient problems were due to the generally low levels of P and K. One or two farms also had water or soils which were somewhat saline or had other mineral deficiencies.

Table 36. Parameters potentially related to yield on the 12 farms, arranged by order of yield on patches to which 120kg N/ha/cut was added

Farm	Total N kg/ha	No of cuts	Response to N (% farm yield)	Water plot 1 (mm)	Water Plot 2 (mm)	Water Plot 3 (mm)	Water mean (mm)	Water EC	Yield Farm	Yield + N120
SL06	360	3	20	724	1,487	414	908	1,700	4.1	4.9
WQ6	600	5	-10	1,434	1,257	703	1,014	1,450	9.6	8.6
BS02	840	7	-19	643	1,579	297	920	1,700	11.8	9.5
DD5	720	6	21	1,486	852	0	1,169	1,100	14.7	17.8
BS03	840	7	9	1618	677	846	1,072	1,800	16.4	17.9
SL02	360	3	79	1,4493	4,258	722	5,667	630	11.2	20.0
SL14	360	3	268	6262	3,616	6,880	3,951	630	6.0	22.1
KH11	840	7	9	1,987	972	293	1,278	1,100	26.3	28.7
DD2	720	6	25	1,866	1,823	726	1,676	2,400	24.4	30.6
NZ04	840	7	27	4,623	3,023	0	3,823	600	27.1	34.3
SM2	840	7	5	2,955	1,222	1,391	1,654	2,200	34.7	36.4
SM3	840	7	22	3,676	1,695	866	2,516	2,700	43.4	53.0
Mean			(24)	3,481	1,872	1,314	2,137	1,501	19.1	23.7

Source: Data from Stephens, 1993b

Table 36 lists some of the parameters, as recorded by Stephens (1993b), which might be associated with yield.

Using date from this table, Figure 24 illustrates the great divergence of mean water application rates on each of the 12 farms, from 908mm to 5,667mm, and the equally great differences in application rates on different parts of the same field (plots 1-3). On average plots 1 received 2.6 times more water than plots 3. In practice, it is very clear, the sprinkler systems were being used extremely inefficiently.

---

[6] Dry Matter.

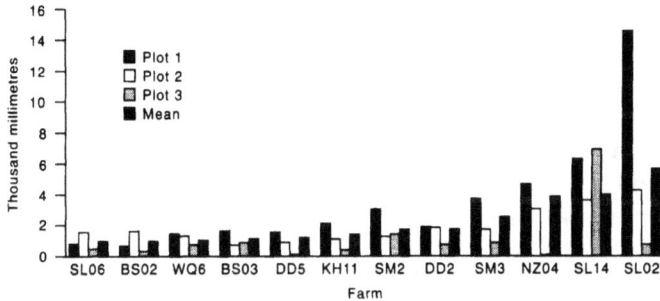

Source: Data from Stephens, 1993b

Figure 24. Total application of water on the 12 farms, and on each plot, arranged by mean value

The differences in application rates were important. A test of the correlation between yields and irrigation depths (Figure 25) was very significant up to the application rates needed to fulfil the theoretical crop requirements, which were calculated at 2,650mm for Rumais (on the Batina coast) and 3,030mm at Nizwa (in the less humid interior). It is interesting that the three farms irrigating considerably in excess of this level (Figure 25) were all using water with very low electrical conductivity. The combination of high rates and low EC will have leached nitrogen (and other soluble salts) from the soil. So, it is not surprising that these three farms showed a very high DM response to the additional nitrogen, resulting in a 72% increase in yield, in comparison with the overall response to nitrogen by the 12 farms combined of 24%.

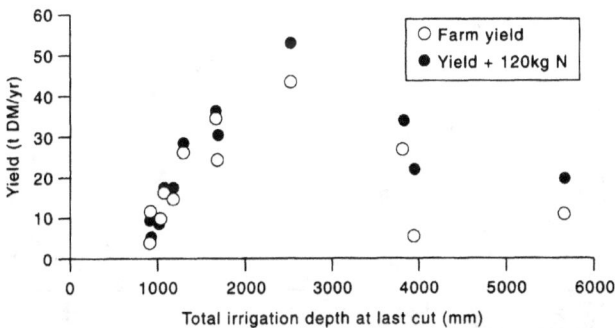

Source: Data from Stephens, 1993b

Figure 25. Total dry matter yield versus total irrigation depth on 12 GMDP farms

Much higher irrigation rates should, obviously, have been used in summer than in winter. The farmers were aware of this and responded to the surveys by saying they had distinct summer and winter schedules, as recommended. However, Stephens (1993b) found that in practice; 'many farms had a relatively constant irrigation schedule regardless of the season' with variations due to external factors 'and not as a conscious decision to increase or decrease the application depth' (p. 24). But did the obvious inefficiency of water use result in significant yield losses and water wastage?

Table 37. 3-farm running means for water application rates and yield means, and absolute and % yield increase

Farm	Water (running mean) (mm) A	Farmer yield (running mean) B	Yield + 120N (running mean) C	Yield (absolute increase) C - B D	Yield (%. increase) (C - B) x 100 E
SL06					
BS02	947	8.5	7.67	-0.83	-9.8
WQ6	1002	12.6	12.00	-0.6	-4.8
BS03	1085	13.57	14.77	1.2	8.8
DD5	1173	19.13	21.47	2.34	12.2
KHI1	1367	25.23	27.63	2.4	9.5
SM2	1536	28.47	31.9	3.43	12.0
DD2	1949	34.17	40.00	5.83	17.1
SM3	2672	31.63	39.3	7.67	24.2
NZ04	3430	25.5	36.47	10.97	43.0
SL14	4480	14.7	25.47	10.77	73.3
SL02					

Source: Data from Stephens, 1993b

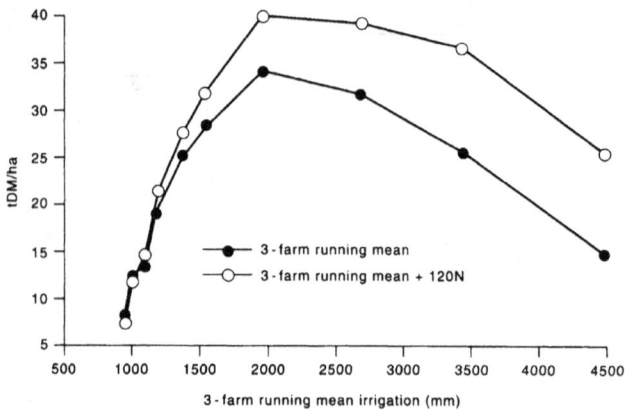

Source: Data from Stephens, 1993b

Figure 26. 3-farm running means, irrigation rates versus yield

Table 37 records the 3-farm running means for water application on the 12 farms. Perhaps the most important point revealed by the table is the rapid percentage increase in yield on the patches given extra N (column E) as shortage of water ceases to be a limiting factor and then again as excess water reduces soil fertility by leaching nitrates.

This is clearly illustrated in Figure 26 which equally clearly shows that excess irrigation is not only a waste of water but also has a strong negative impact on yield.

Table 38. 5-farm running means of irrigation rates and yields

Water mean (mm)	Water 5-farm running mean (mm)	Farmer yield	Farmer yield 5-farm running mean (t DM/ha)	Yield +120N	Yield + 120N 5-farm running mean (t DM/ha)
908		4.1		4.9	
920		11.8		9.5	
1,014	1,017	9.6	11	8.6	12
1,072	1,091	16.4	16	17.9	17
1,169	1,237	14.7	20	17.8	22
1,278	1,370	26.3	23	28.7	26
1,654	1,659	34.7	29	36.4	33
1,676	2,189	24.4	31	30.6	37
2,516	2,724	43.4	27	53	35
3,823	3,527	27.1	22	34.3	32
3,951		6		22.1	
5,667		11.2		20	

Source: Data from Stephens, 1993b

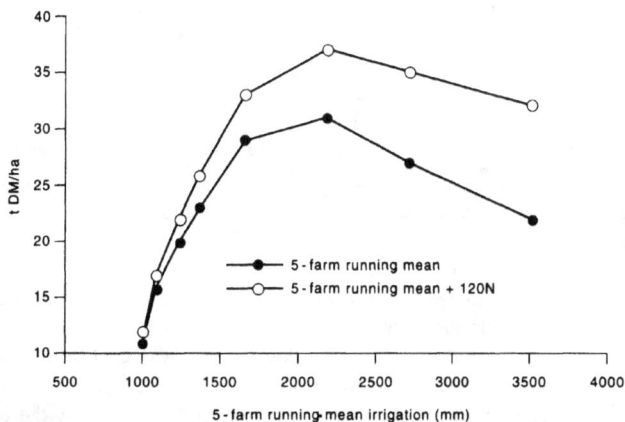

Source: Data from Stephens, 1993b

Figure 27. 5-farm running means, irrigation rates versus yield

The additional smoothing effect produced by the 5-farm running means of irrigation rates and yields (Table 38 and Figure 27) indicate that peak yields of 31t DM/ha (farmers' patches) and 37t DM/ha (patches with extra 120N) were produced with an irrigation rate of about 2,200mm, which was only just over the mean irrigation rate for all 12 farms and all plots (Table 36). Therefore, in theory, with no additional use of water (only a better distribution of it), major yield increases would have been possible (Table 39). If the farmers had used more nitrogen and an ideal rate of water application the yields per hectare would have risen from 19.1 tonnes to 37 tonnes, a massive 93.7% increase.

Table 39. Rhodes grass yield variation, with N and assuming ideal water distribution

| | Yields: t DM/ha | | |
	With actual water distribution	With theoretical water distribution	Increase in yield
Farmer's field	19.1	31	11.9
Farmer + 120N/cut	23.7	37	13.3

Source: Data from Stephens, 1993b

Figure 28 indicates that there is very little if any correlation between yield and EC of the water.

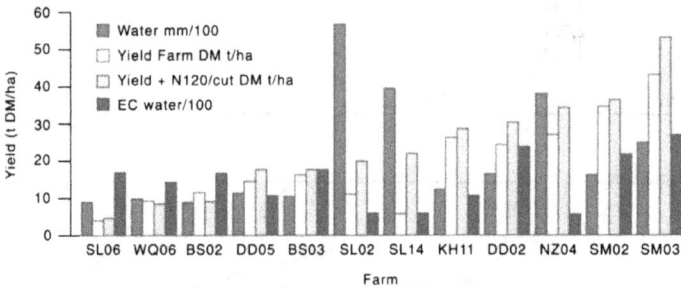

Source: Data from Stephens, 1993b

Figure 28. Rhodes grass yields on the 12 farms, compared with water EC and application rate, by yield on patches to which 120kg N/ha/cut was added

This lack of correlation is made clearer in Figure 29, and this in spite of the fact that EC's varied from as low as 630 up to 2,700. The higher levels would normally be considered too saline for the irrigation of most crops, but in fact some of the highest yields were associated with the highest EC values. Rhodes is tolerant of saline soils and water.

But Figure 29 also indicates that in general there was a close relationship between the yields without and with additional nitrogen. In most cases the added nitrogen resulted in an increase in yield, with proportionally the biggest increases on farms SL02 and SL14 where the irrigation rates were very high (and EC very low). On the other hand, on the farms which applied very low levels of irrigation water the impact of additional nitrogen was low or even negative because the soil was not sufficiently moist to allow the nitrogen to be utilised. On several farms when irrigation was inadequate Stephens (1993b) observed wilting in the fertilised patches before the unfertilised patches. This was probably due to the expansion in leaf area in response to nitrogen and the plant's inability to sustain transpiration from it.

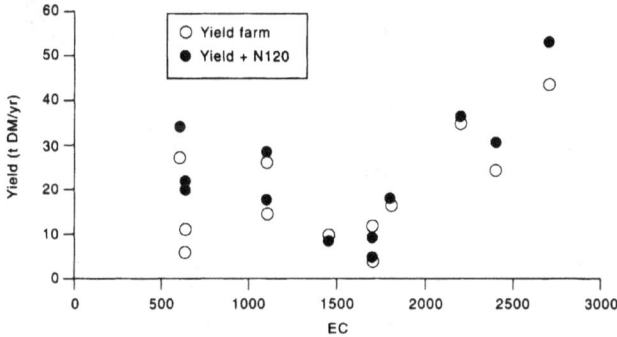

Source: Data from Stephens, 1993b

Figure 29. Total dry matter yield versus EC of irrigation water

Table 40. 5-farm running means for cuts and/or N application and yields

Total Nkg/ha	5-farm running mean	Farmer yields	Farmer yields 5-farm running mean	Yield +N120	Yield +N120 5-farm running mean
360		4.1		4.9	
360		6.0		22.1	
360	480	11.2	9.12	20.0	14.68
600	552	9.6	13.18	8.6	19.82
720	648	14.7	14.34	17.8	17.30
720	744	24.4	15.38	30.6	16.88
840	792	11.8	18.72	9.5	20.90
840	816	16.4	22.72	17.9	24.62
840	840	26.3	26.52	28.7	29.10
840	840	34.7	29.58	36.4	34.06
840		43.4		53.0	
840		27.1		34.3	

Source: Data from Stephens, 1993b

The final point to emerge from Stephens data is that the 5-farm running mean of extra N application versus DM yield (Table 40 and Figure 30) indicates that there is not only a response to additional N but also that the highest rates of N give the highest yields. However, because N was added to the patch to coincide with every cut made by the farmer and because the number of cuts varied from three to seven, the increase in yield may have been partly due to cutting the grass more frequently. In fact this is suggested by the fact that the farmers yields also rose sharply.

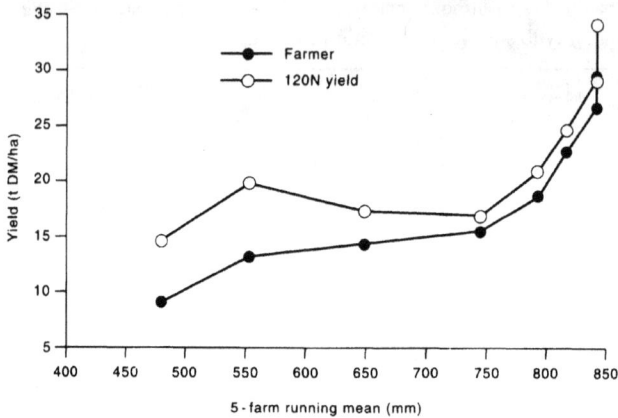

Source: Data from Stephens, 1993b

Figure 30. 5-farm running means, nitrogen/cut versus DM yield

Unfortunately there were also other reasons for poor yields of which the three most important were: other crop irrigation priorities; harvesting methods; lack of understanding of hay-making.

The significance of other crop irrigation priorities was revealed on farm NZ04 when it was seen that completely contrary to the recommended schedule, much more water was being applied in winter than in summer. A process of careful questioning and calculation revealed the explanation. The 360 *gelbas* that the farmer had under flood irrigation - mainly alfalfa - took so much time during the working day in the summer months that little time was left to operate the sprinklers on the Rhodes grass!

For Rhodes productivity, differences in harvesting methods were shown to be very important. The most common harvesting method was by hand, using the serrated edge sickle traditionally used to cut alfalfa. We had learned in al-Khabura (and previously during our survey work in the Dhahira) that farmers universally believed that cutting alfalfa very close to ground level was important to ensure good re-growth. They therefore adopted the same approach to cutting Rhodes grass, which left discrete clumps of established grass but removed all the stolons that would otherwise have colonised the bare ground. The space between the clumps was easily invaded by weed species taking advantage of the lack of competition. Cutting Rhodes by sickle tended also to mean that it was cut in irregular patches, sufficient only for immediate needs. This made the task of operating an effective management regime for irrigation and fertiliser application almost impossible. On the other hand, as a result of the long experience we had acquired on the Khabura Project with pedestrian, powered, reciprocating blade cutters, MAF started to encourage the use of these machines on the GMDP farms. By 1993 farm NZ04 was one of a minority using such a machine. In practice large areas were cut at one time several centimetres above ground level therefore leaving all the stolons and creating a dense sward which was higher yielding and difficult for weeds to penetrate. In the 12-farm study, the five farms with the highest yields included the two which used a powered cutter and two which used a less severe hand cutting technique. The remaining farmers all cut their Rhodes as they cut their alfalfa.

Making hay is important in the annual management cycle of Rhodes grass. Because of the seasonal nature of Rhodes production, forage self-sufficiency for the goats depended on maximising yields in summer, well above summer requirements, so that the surplus could be converted to hay for winter feeding, as practised on the Khabura Project's farm. In practice, however, few GMDP farmers made any hay because the management of the irrigation during the critical summer months did not allow the full yield potential to be attained. Also, they will not have had the incentive because haymaking was an entirely new idea in northern Oman at that time, except for those farmers influenced by the Khabura Project and a few commercial farmers on the Batina coast.

Table 41. Animal numbers, Rhodes grass area, farm DM production and requirements

Farm	Area (ha)	Animal units	DM req (t/yr)	DM prod (t/yr)	Prod as % required
KH11	0.6	8.375	22.9	15.9	69.4
NZ04	0.43	7.875	21.5	11.7	54.4
SL14	0.4	6.4	17.5	2.4	13.7
SL02	0.27	6.3	17.2	3.0	17.4
DD2[1]	0.38	5.8	15.8	9.1	57.8
SM2	0.36	5.0	13.8	15.6	113.9
SM3	0.32	4.7	13.7	11.0	85.9
DD5	0.12	3.35	9.1	1.7	18.7
BS3	0.20	3.275	8.9	3.3	37.1
SL06[1]	0.39	2.875	7.8	2.4	30.8
BS02	0.22	2.55	7.0	2.6	37.1
WQ6	0.18	2.25	6.1	1.8	29.5
MEAN	0.32	4.896	13.44	6.71	(49.9)

Source: Data from Stephens, 1993b

(1)   Includes some production from flood irrigated areas

Low yields impacted on livestock numbers. By analysis of the data in Table 41 Stephens showed that there was a significant correlation (at the 1% level) between the sprinkler irrigated area of Rhodes, as well as DM production (at the 5% level), and the total number of 'animal units' on the farm, that is to say the goats in the scheme and the goats and other livestock also owned by the farmer. The percentage of DM requirements being satisfied by production was very variable, ranging from 14% to 113%, with a mean of only 50%. Stephens concluded that stock numbers were being limited by the level of Rhodes production, rather than the other way round.

Stephens (1993b), as a result of his detailed studies, made a set of specific technical recommendations which are worth recording in full:

Farm selection studies should include:

- estimation of peak demands based on the existing cropping patterns, irrigation systems, efficiencies, management, soil types and climate;
- measurement of the well yield and capacity of the pumpset to power a sprinkler system;
- measurement of the impact that a sprinkler system would make on the total water use.

All selected farms should then:

- undergo a detailed evaluation of cropping patterns, current water use, and irrigation systems;
- be given specific recommendations on ways of improving the efficiency of water use;
- be encouraged to substitute part of the existing irrigated area with the sprinkler irrigated Rhodes grass.

Farmers/labourers who operate the system must learn the importance of using:

- correct pressures;
- correct schedules to apply the correct depth of water at the appropriate intervals - it is technically very feasible to provide a schedule specific to the circumstances of each farm;
- correct maintenance to keep all parts of the system in good condition.

With the water delivery system then under control, there are three main concerns for the husbandry of the Rhodes grass:

- fertiliser use: the rates and intervals for the application of nitrogen, potassium and phosphorus are of particular importance;
- cutting method: either use a mechanical cutter, or at least ensure that hand cutting does not remove the stolons and is high enough to encourage a dense sward.
- hay making: the small farmers must understand the techniques for making hay (one to two days drying in the summer sun is sufficient), have storage capacity for it and understand its value as a winter feed when fresh forage is in short supply. Only then will the farmers have the incentive to maximise peak season summer production and productivity above their summer feed requirements.

Monitoring, evaluation and extension:

- the monitoring procedures used by Stephens need to be maintained;
- the data need to be evaluated so that meaningful conclusions and helpful recommendations can be made;
- the extension service needs to be trained in both monitoring and evaluation, and trained also in the complex skills of working with the farmers in ways which will give them the confidence and interest to heed the extension advice.

In summary, it was theoretically possible to have almost doubled farmer yields from 19.1t DM/ha to 37.0t DM/ha by a better distribution of water on the same area of land, and using the additional N application rates followed by Stephens. It also seems that the yields could have been further increased by more frequent cutting and/or more frequent applications of fertiliser. And of course an additional huge increase in production would have been possible if the proportion of land under crop for which capital equipment had been installed had risen from the actual 22% to the possible 100%. The use of mechanical harvesters and a knowledge of haymaking would both also have led to increased Rhodes production.

However, motivation is as essential as having a good knowledge of improved techniques. It is all too common for extension services to be caught in the awkward position of having to 'push' farmers into adopting a technique against their will. Extension services are likely to function much more successfully if the farmers are first motivated to succeed, and then turn to the extension service for advice. Often the main motivation stems from the fact that the farmer and his family have no alternative source of income or way of life. However, in Oman and the Gulf, with their oil wealth, this source of motivation is now far from dependable; in recent decades people have been able to find alternative income sources by leaving their villages. Therefore the income from the goat scheme has to be positive and large enough to form a significant proportion of total family income, and the labour involved must not be too arduous. People must feel convinced, during the learning phases of the scheme, that the effort to master the new skills is going to be worthwhile and in the long-term interests of themselves and their families. The main role for the government is to establish the economic context in which real profits are feasible for at least the better farmers. This may involve output subsidies or it may involve restricting the import of live sheep and red meat so that the market price of the local animals is high. If the farmers are, as a result, making a profit it will make the extension centres' task of optimising efficiency of water use and Rhodes grass production much easier.

5.3.3 Goat management and productivity

Taylor (1991b) devised a computerised monitoring system (Ch. 5.3.6) which helped ensure that the data being collected from the GMDP farms were duly recorded. Periodic reports on goat productivity and problems were then returned to the extension centres together with recommendations for enhancing the data collection system and for helping the farmers ameliorate problems that the data analyses had revealed.

Table 42. Results from the 40 monitored farms from the GMDP, 1988/9-1990/1

Item	1/12/88-31/8/89		1/9/89-31/8/90		1/9/90-31/7/91	
	Project[2]	All[3]	Project[3]	All[3]	Project[2]	All[3]
Av. no adult does present[1]	197	285	349	579	466	747
Total kiddings	203	247	331	461	322	400
Total kids born	232	292	394	551	387	482
Mean litter size	1.14	1.18	1.19	1.20	1.20	1.21
Kiddings/100 does	103.05	86.67	94.84	79.62	69.10	53.55
Kids/100 does	117.77	102.46	112.89	95.16	83.05	64.52
Av days since last kidding	193	201	280	281	364	368
for (n) kiddings this period	3	5	175	203	242	285
% kids born this period:						
(a) died to date	10.5	10.4	11.1	12.4	7.1	6.6
(b) sold/slaughtered to date	36.8	37.8	36.0	33.8	6.1	7.9

Source: Taylor, 1991d

(1) Any doe brought from outside OR over 450 days old (av. no. = total animal days/days in period
(2) 'Project' refers to goats originally given by MAF PLUS their offspring
(3) 'All' refers to the total goats on the farm - many farmers owned non-project goats as well

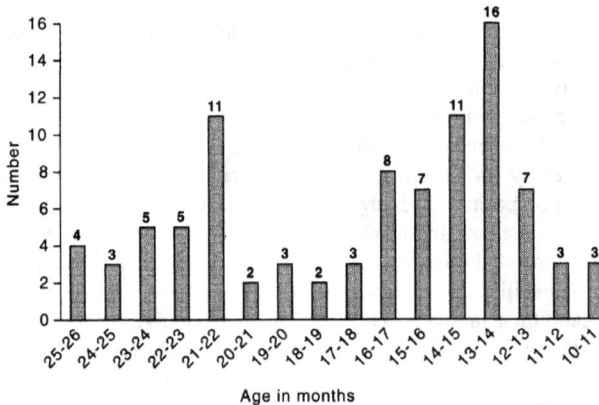

(1) 93 first kiddings

Source: Taylor, 1991d

Figure 31. Frequency distribution of age at first kidding for GMDP goats[1]

Data from the 15% of the farms most intensively monitored (46 of 300), some in each region of northern Oman, were entered into the system and at the end of the third year (end-July 1991) were fully analysed (Taylor, 1991d). The first screening of the data eliminated all five farms from the North Batina region and one other farm because the quality of the raw data was poor. The key main findings from the other 40 farms are recorded in Table 42.

Although the average number of adult does on the GMDP farms continued to rise steadily, the breeding performance was becoming 'sub-optimal' as time passed as a result of poor (or worsening) performance in a number of key parameters: age at first conception, kidding interval, litter size and growth rates to 90 and 365 days - as detailed below.

Clearly, little attempt was being made to manage the date of first conception/kidding though initially this did not seem to be having serious repercussions. Batch breeding at the Khabura Project had allowed the age at first conception to be brought down from 523 days to 344 days for cross-bred goats, the largest group. It also came down from 666 to 468 days for the local goats by which age the standardised weight of the local goats was about 28kg. In comparison, the age at first conception of the GMDP local goats (up to August 1991) was about 366 days when the standardised weight was about 25kg (Taylor, 1991d; Addy and Taylor, 1991); not very different from the Khabura Project's cross-breds. The mixed record of age at first kidding is illustrated in Figure 31, which excludes the 5% of goats that had not kidded at all. Some of the project females were too young when they first kidded, while one-third were too old, at 600 days or older.

Table 43. Kidding intervals of local goats on monitored GMDP farms, and on the Khabura Project

	Mean interval (days)	Nos. intervals	Range
Recorded kidding 1st to 2nd	323	297	169-722
Recorded kidding 2nd to 3rd	307	83	172-455
Farmer does	331	64	190-676
Project does	316	318	169-722
Overall	318	385	169-722
Khabura (not batched)	281	59	
Khabura (batched)	257	150	

Source: Taylor, 1990a and 1991d

For subsequent kiddings the GMDP goat kidding intervals were considerably greater than on the Khabura Project (Table 43), though this may simply have reflected the difference in target intervals; 243 days (8 months) in the latter and 365 in the former.

In August 1991 a study of the most recent kiddings of the 326 original does that were still on the GMDP farms revealed the following:

> 70% (229) had kidded within the last 12 months
> 19% (63) had last kidded 12-24 months previously
> 6% (18) had last kidded over 24 months previously
> 5% (16) had not kidded at all

Thus the breeding performance was indeed sub-optimal. The fact that 30% of the does had not kidded within the previous 12 months was a cause for concern. The 11% that had never kidded or not kidded within the past two years should probably have been culled.

Table 44. Litter sizes of goats on monitored GMDP farms

	Mean kid nos.	Nos. litters	Range
1st recorded kidding	1.14	699	1-3
2nd recorded kidding	1.22	298	1-2
3rd recorded kidding	1.31	83	1-3
Farmer does	1.18	266	1-2
Project does	1.18	761	1-3
Project does daughters	1.04	46	1-2
Overall	1.17	1084	1-3

Source: Taylor, 1991d

The mean litter size (Table 44) of the goats on the GMDP farms at 1.17 was 7% less than on the Khabura farm, which was 1.26 for local goats.

Table 45. 365-day weights of goats on monitored GMDP farms, adjusted to a sex ratio of 50M:50F

Origin of dam	Mean 365-day wt (kg)	Nos. kids
The farmer	21.6	43
GMDP	25.3	154

Source: Taylor, 1991d

Kid growth rates also were slower: a standardised 365-day weight of 25.3kg for the GMDP goats (Table 45) compared with 28.6kg for the local goats on the Khabura Project (Table 1).

Table 46. 90-day weights of goats on monitored GMDP farms, adjusted to a sex ratio of 50M:50F

Project year of birth	Litter type	Mean 90-day wt (kg)	Nos. kids
1	Single	13.7	184
1	Multiple	12.8	80
2	Single	11.7	290
2	Multiple	10.8	128
3	Single	10.9	59
3	Multiple	10.1	23

Source: Taylor, 1991d

In 1991 it was possible to check 'year' effects for the three full years of the GMDPs life. Kiddings per 100 does and total kids per 100 does had fallen each year, from 103 to 69 and from 118 to 83 respectively (Table 42) - contrasting with means of 100 and 175 for all Khabura Project goats in 1987 and 1988 (Table 18). Weight comparisons could also be made each year for kids up to 90 days. At that age, kids born in the second season of GMDP were 1.9kg lighter than kids born in the first season, while kids born in the third season were 0.8kg lighter than kids born in the second season (Table 46). These differences were statistically highly significant, and also very disturbing because of the steep downward trend in kidding rate as well as weight gains.

In practice no serious attempt was made to batch the kiddings on the GMDP farms, although the new pens created this possibility. In the first year because of the timing of mating in the period before the majority of the animals were distributed, most of the kiddings were in the March to May period. But in the second and third years an annual rhythm asserted itself with most kids born in the months November to March. We can accept that any attempt to enforce an 8-9 month kidding cycle would probably have been unwise, given the lack of experience of both the farmers and the extension services, and the conclusions from the Khabura Project (Taylor, 1990a). However, the Khabura Project had also shown the very important management advantages of batch breeding (reflected in growth rates of the kids), so it is unfortunate that the kids on the GMDP farms were born over such long time intervals. Managing the nutrition of the kids and the does will have been virtually impossible.

Table 47. Offtake and deaths of kids on monitored GMDP farms[1]

Means of disposal	Total born	Died (n)	Rate (%)	Sold (n)	Rate (%)	Home slaughter (n)	Rate (%)
Born year 1	389	28	7.2	36	9.3	11	2.8
Born year 2	641	43	6.7	39	6.1	49	7.6
Born to farmer doe	272	26	9.6	31	272	11	4.0
Born to project doe	727	41	5.6	44	6.1	49	6.7
Single	723	44	6.1	-	-	-	-
Multiple	307	26	8.5	-	-	-	-
Male	515	-	-	58	11.3	44	8.5
Female	515	-	-	17	3.3	16	3.1
Overall	1030	71	6.9	75	7.3	60	5.8

Source: Taylor, 1991d

(1) Adjusted to a sex ratio of 50M:50F

Table 48. Herd composition on the monitored GMDP farms

Item	1/12/88-31/8/89		1/9/89-31/8/90		1/9/90-31/7/91	
	Project	All	Project	All	Project	All
FEMALES						
< 90 days	30	36	28	42	18	25
91-450 days	82	99	161	213	172	220
over 450 days	306	457	420	691	543	849
MALES (inc. castrates)						
< 90 days	25	33	24	30	15	16
91-450 days	74	95	141	200	152	186
over 450 days	0	5	39	59	80	126
Total	517	725	813	1,235	980	1,422

Source: Taylor, 1991d

Recorded death rates were low (Table 47), but the total recorded offtake and deaths at only about 20% (206 of the 1,030 kids) were improbably low even allowing for all the female kids born being retained to increase herd size.

211

However, part of the explanation for this is that the farmers were keeping both male and female offspring on the farm in order to increase herd size, rather than selling most male kids for meat (Table 48). Many more females than males should have been retained. The fact that so many males were kept on-farm suggests that marketing strategies could have been improved.

The value of the above analyses is not only that they showed mean productivity to be poor, but also that they pin-pointed the three principal contributory factors, namely worsening kidding rates, increasing intervals between kiddings and declining kid growth rates, all made worse by poor marketing strategies. Because of the analyses, it then became possible, at least in principle, for the Animal Production Officers to identify the causes of the problems and for the extension staff to bring the problems and the solutions to the attention of the farming community.

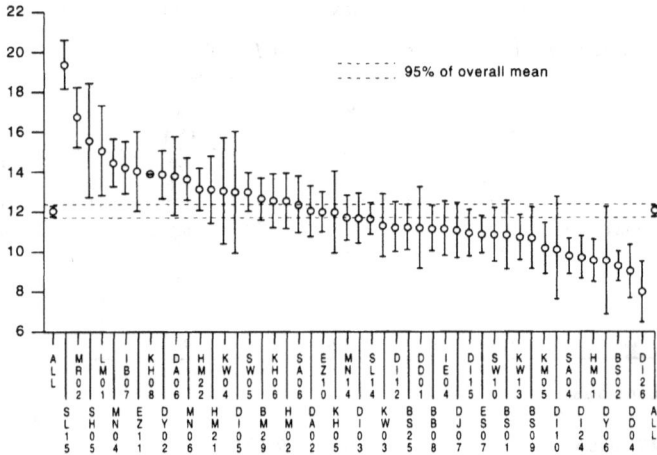

Source: Taylor, 1991c

Figure 32. Mean 90-day weights with 95% confidence limits for individual farms (kg)

It is also important in the context of livestock extension that the performance of the different animals and farms was not uniformly poor. Some 102 of the original does had, for example, kidded three times with a mean kidding interval of only 322 days. Also Taylor (1991c) clearly showed that some farms greatly out-performed others in terms of mean daily weight gains of their kids up to 90 days and 365 days (Figure 32). Unfortunately no attempt was made by the extension service to isolate the reasons for these inter-farm variations. They were probably related to management practices which, if identified, would have been very important aids to the effective delivery of extension advice.

5.3.4 Veterinary support

An objective of the goat and sheep breeding work based at ABARC was to increase the mean genetic potential of different breeds of goats and sheep in Oman, and provide the improved genetic stock to the 'leader farmers' in the GMDP scheme. In practice, there were differing views about who should receive the improved animals. On the one hand, some people in MAF wanted to leave the question open so that any interested farmer might be eligible. On the other hand, it was argued that the leader farmers on the GMDP scheme should be the primary recipients, in part because we were monitoring their initial progress which was showing which farmers were truly more interested and more able. We felt that such farmers should be rewarded for their efforts and expertise, and that this would encourage the others. A member of the Durham University team particularly concerned about the quality of small ruminant husbandry on the GMDP farms was Louis Kwantes, the veterinary officer on ABARC, who was maintaining a high level of veterinary health amongst the sheep and goats under his control. It was feared that if such animals were then placed on GMDP farms which had a poor health status, their performance was liable to suffer severely, and their genetic potential not be attainable. Therefore, during 1992 Kwantes (1993) undertook a survey to assess the nature of veterinary work among the GMDP farms, and characterise the disease problems encountered on them (Kwantes, 1993). He visited each of the 25 clinics servicing the GMDP scheme and, with the staff, completed an initial questionnaire. The staff were then asked to complete a short report form for each disease treatment undertaken on a GMDP animal throughout 1992. In practice 576 forms were usable in the analysis after discarding 87 forms that were inadequately completed in one respect or another. The data were analysed for type of disease and number of cases. Significant diseases were defined as comprising more than 10% of all cases in a centre, or more than 4% in a region, and more than 1.1% overall. Some 48 conditions were recorded, divided into the categories: digestive, infectious, parasitic, reproductive/genitalia, and other. Overall, the significant problems were as recorded in Table 49.

The parasite problem may be under-represented in Table 49 because when it was recorded together with a more serious problem only the more serious problem was listed.

There were some regional variations which may reflect differences in quality or priorities of the veterinary services in different centres. For example, parasites were a significant problem in Oman Dakhil (Nizwa and Izki). Also, whereas contagious diseases were largely held in check on the GMDP farms, CCCP formed 8% of all diagnoses in Oman Dakhil, while foot and mouth disease (FMD) totalled 23% of cases in the Sharqiya (though all from one outbreak).

Seasonally, pneumonia and upper respiratory tract problems followed a marked pattern with peaks in November-January and May-July, particularly on the Batina. Kwantes (1993) believed this pattern to have been correlated with the change in seasons from cool to hot and hot to cool; temperature changes are commonly seen as stress factors in the occurrence of disease. Seasonal patterns for other conditions, such as diarrhoea/enteritis and intestinal parasites were less clear.

Table 49. Significant veterinary problems recorded in the 1992 survey of GMDP farms

Disease	Cases (n)	Cases (%)	Comparison	
			Khabura vet. clinic (%)	ABARC (%)
Pneumonia and/or upper respiratory tract	501	28	(43)	2
Diarrhoea and/or enteritis	394	22	8	26
External parasites	314	17	3	
Intestinal worms	89	5		
Contagious caprine pleuro-pneumonia	81	4	(43)	
Copper deficiency and/or paralysis	71	4	2	
Foot and mouth disease	57	3		
Mastitis	47	3	4	7
Other (1% or less each)	253	14		
TOTAL	1807	100		

Source: Kwantes, 1993

It is instructive to compare the GMDP farms and the Khabura Project clinic farms with ABARC where the breeding stock were very effectively isolated from other animals and where veterinary care was given a very high priority (Table 49). The major difference concerns pneumonia and upper respiratory tract diseases, accounting for 28% of clinical cases on GMDP farms and 43% at the Khabura clinic but only 2% at ABARC where 'it is not considered a problem at all' (Kwantes, 1993, p. 29). Part of the explanation lies in the lack of isolation of animals on the GMDP farms, and part with the regular use of a pasteurella vaccine at ABARC.

Perhaps the most important findings of the survey concerned not so much individual disease diagnoses, or the use of particular vaccines and other medication, but rather the overall management and delivery of the service. A major point is the relationship between the clinician and the farmer. Only 15% of the treatments were performed in the clinics but of these in 74% of the cases the owner was in attendance. A further 19% of treatments were made on the farm at the request of the owner, and of these 70% of the owners were directly involved. However, 61% of all consultations were made on routine farm visits, and in these cases only 19% had owner participation. This relative lack of farmer participation minimised the likelihood of proper follow-up treatment being administered and made the veterinarians task of disseminating information and extension training almost impossible. Indeed, the veterinary staff felt that one of their two major problems in providing quality care to GMDP animals was communication with and co-operation from the owners.

Significantly, the survey also showed that in most cases there was little or no difference between the disease problem as suggested by the farmer and as diagnosed by the veterinarian, which indicates, to say the least, that the diagnostic capabilities of the veterinary staff were not being put to optimum use. The treatment administered also was sometimes not dictated by drug of choice but by drug availability; the range of medicines for dispensing was very limited. Almost without exception the advice given and follow-up were inadequate. Post-mortems were rarely undertaken (often dead animals were not even seen by the veterinary staff), and laboratory facilities were widely under-utilised. Many of the laboratory diagnoses were of doubtful clinical importance, or of no consequence to the treatment regime, or not even requested in the first place. Most of the time the only 'follow-up' was that the animal was not seen again and therefore presumed fully recovered! The morale of the veterinarians would have been improved if more active interest had been taken in their work, and if a proper programme of in-service training had been instituted.

However, one reason for making the veterinary survey was to highlight the suspected problems and recommend way of ameliorating them. In summary, in somewhat modified form, the following are the key recommendations that Kwantes (1993) suggested:

- continue the monitoring programme to identify significant disease conditions and any seasonal and sub-regional patterns of occurrence;
- develop protocols for interventions;
- develop an in-service training programme for the veterinarians and create the context in which they use their skills and the diagnostic and other laboratory facilities available to them to the full;
- address the problem of poor communication and lack of co-operation with farmers - spending more quality contact time with responsive farmers;
- investigate major diseases more thoroughly, notably the pneumonia and respiratory tract problems;
- ensure that proper medication and equipment are available to the veterinarians, including easier access to field transport.

Unfortunately, the quality of the veterinary services was inadequate to the task. It did not have the capability to cope with a major disease outbreak, should one have occurred.

5.3.5 Profitability

Of the 300 farms in the GMDP, a detailed financial analysis was made of a cross section of 40 of them in 1991 (Conroy, 1992c-d). Concerning returns, sale of goats and consumption by the farm household were the main benefits that the owners obtained from these farms, though some additional benefit was derived from the use of goat dung as a fertiliser. On the costs side of the equation, the main variable costs were for feed, with roughly equal amounts being spent on fodder and concentrates (Figure 33). Most fodder was grown on the farm, but

16 of the 40 farmers bought some fodder including six who bought more than half that their animals consumed. Concentrates were the single largest input cost, accounting for 55% of all variable costs. High levels of fodder and concentrate purchases suggest that such farms will have been in financial deficit because Taylor (1991e) had already shown that livestock farms were most unlikely to be profitable if these feedstuffs were bought in significant quantities.

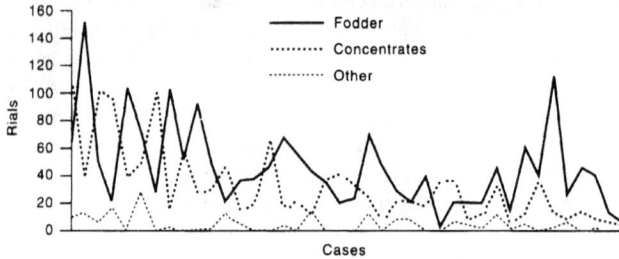

Source: data from Conroy, 1992c

Figure 33. Fodder, concentrates and other variable costs per breeding doe

Conroy (1992c) estimated the mean gross revenue per breeding doe at RO42.3. However, the variable costs of production (concentrates, seed, fertiliser and irrigation, etc) on 80% of the farms exceeded the gross revenue, and these farms therefore had negative gross margins. In fact, the mean gross margin of all the farm was negative, at RO-40.3/breeding doe, though six farms had positive gross margins. The mean fixed cost per breeding doe was RO39.6 of which RO33.2 was labour. Only two farms had small positive net margins after the fixed costs had been taken into account. Although the mean performance was so poor, half the farmers stated that they thought their goat enterprises were profitable.

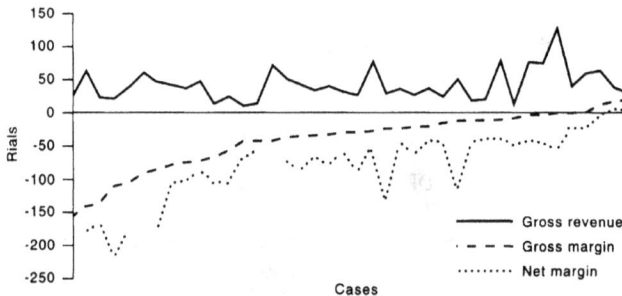

Source: date from Conroy, 1992c

Figure 34. Gross revenue, gross margin and net margin per breeding doe

In Figure 34 the 40 cases analysed are arranged in order of gross margin per breeding doe, from the greatest loss of more than RO150/doe to the half dozen which recorded a small positive value. There is no correlation with gross revenue but comparison with Figure 33 shows that there is some correlation between low gross margins and very high fodder and concentrate costs.

If one is trying to find room for optimism in the two Figures it lies in the extreme range of gross margins and the equally highly variable feed costs. If some farmers can achieve a positive result it should be possible for the extension services to learn from their experience and improve the productivity of the others. As has already been shown, there was enormous room for improvement in all aspects of the production system: irrigation, forage and livestock.

5.3.6 Monitoring

One of our tasks for the GMDP, having undertaken a number of monitoring and evaluation exercises of parts of the scheme as noted above, was to recommend how the collection and analysis of relevant data could be systematised and made on-going. This task was undertaken by Conroy (1992a) who started by reporting on the monitoring activities then in place, and made recommendations to make them more effective.

The main reasons for collecting data on input costs, and on output revenue are:

- to clarify whether participant farmers are making a profit or a loss on their goat enterprise;
- to identify which farmers have the lowest costs per breeding doe, in order that lessons may be learned from them for the benefit of other farmers;
- to monitor the impact of the extension service - did the actions of its staff make the farms more profitable?

Figure 35 represents the computer heart of the monitoring system - for production and gross revenue of the goats - as devised by Taylor (1991b) using the Panacea database and analytical system. The 'Field Data From Monitored Farms' are entered in the 'Data Entry Data File', which is a temporary handling area for new data, where they are entered, checked and processed before passing to the permanent data files. These are of two types: individual 'Farm Data Files', and 'Summary Data Files'. Some automatic analysis of the data is undertaken, which generates information about litters, growth rates, offtake and mortalities and farm and year comparisons. The file for each farm has complete ID, production and departure information for each animal. The summary files include separate files for each birth, each weighing of the animals, each abortion and each animal removed from the farm. These files also allow doe performance, growth rates, offtake and mortality to be analysed for all the monitored farms as a whole.

The system was devised to be very user friendly. Data are collected in the field on printed forms designed to assist data entry. Entering data into the database requires hardly any knowledge of the PANACEA programme, and special menus guide the operator through each task. As the data are entered a procedure checks for logical errors such as birth weights outside a reasonable range, and a printout prompts the operator to verify the suspect data and correct if necessary. Derived figures such as birth interval, mean litter size, and estimates of 90, 180 and 365 day weights are calculated automatically.

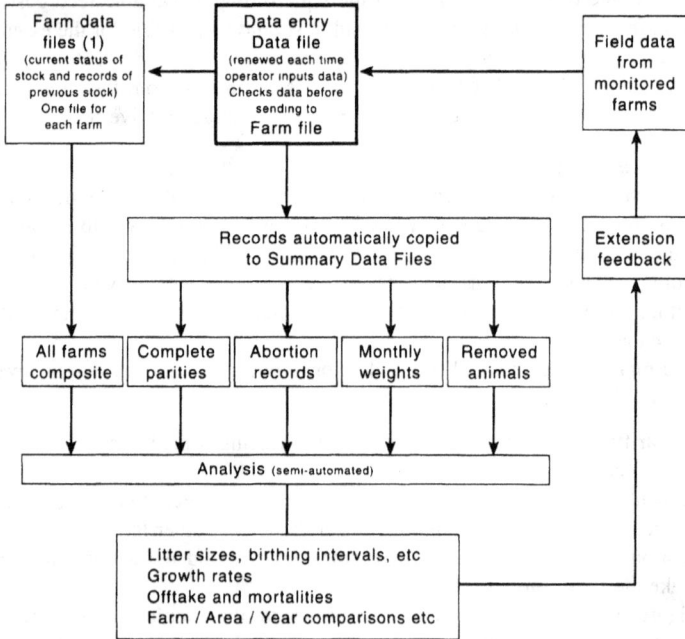

Source: after Taylor, 1991b

Figure 35. Diagrammatic representation of the computerised monitoring system

Perhaps one of the most useful aspects of having the system is to encourage the extension staff to continue collecting the field data with accuracy and enthusiasm because results are fed back to them which allow them to know which farmers require extra assistance to bring them up to standard. The system is equally applicable to sheep and goats.

According to Conroy (1992a) the system generally worked very well, provided that the extension officers in fact made the necessary visits and passed their monitoring forms on to the Department of Animal Wealth for computerisation. The only data that could readily be misreported are the reasons for a goat departing from the farm. A farmer could, for example, say that a goat had died when he had in fact sold it; or he could report a sale price lower than the one that he actually received. In both cases, the farmer's motivation would be to give the Ministry the impression that he is poorer than he really is, in the belief that this would increase his chances of getting additional Ministry grants or subsidies in the future. This kind of attitude was widespread among the farmers interviewed by Conroy, although there is no particular evidence that it led to the misreporting of GMDP production data.

Inaccurate tagging of the kids was another minor inadequacy in implementation of the monitoring system, and extension officers did not always weigh goats up to a year old. Nor were all sale price data recorded which made estimates of gross revenue more difficult to ascertain with accuracy. Unfortunately MAF did not (and does not) have a culture of wanting to check the economic consequences of the schemes it introduces, and so encouragement for the extension staff to undertake long-term economic monitoring was lacking.

Monitoring inputs, by its nature, was much more difficult than monitoring goat production and sales prices. Their costs, as revealed by the financial survey undertaken by Conroy, were unexpectedly and undesirably high. Monitoring was therefore very important in order to see whether their usage could be reduced without affecting productivity; as the price is fairly stable, the monitoring could focus on the quantity fed to the goats. However, data accuracy will be problematic. Conroy felt that greater accuracy would be obtained if the extension officers by direct observation totalled the bags bought and stored at the farm, and marked each bag recorded. He also concluded that monitoring the cost of fodder production and the quantity fed to the goats was too difficult to give meaningful information. The best result would be achieved by calculating the area of land used to grow fodder for the goats, and then to estimate the cost of growing the crop. Differing irrigation costs and efficiencies from farm to farm are the most important element of fodder production costs.

5.4 A national programme but too narrow a vision

At the national level the GMDP or 'leader farmer' scheme embraced some of the aspects of the schemes adopted by the Khabura Project's 'participant farmers' including goat and sheep breeding, use of Rhodes grass, and more intensive livestock management. The GMDP was also, appropriately enough, used by the Minister as a vehicle to promote the use of sprinkler irrigation. In addition, the Minister attempted, through the GMDP to focus some improvements into the livestock extension services. But the national programme's vision was too narrow in that no thought was given to the design or manufacture of local pens, tools and equipment, or to the use of by-products (including milk, wool and dung) or to the development of irrigation systems that could be assembled, installed and maintained by local artesans.

Also, the GMDP expanded too fast and therefore included farmers who were not really interested in it and whose farms did not even have sufficient water.

219

Many of the so-called 'leader farmers' were therefore failing, which made meaningless the whole leader farmer concept.

Worse still, MAF was not very concerned with learning from the experience. With ODA support we were, nevertheless, able carefully to monitor and evaluate the management, by the farmers, of the goat husbandry, irrigation and Rhodes grass production systems. Many weaknesses and failings were thereby revealed, as described above, but the work also showed that a few farmers were succeeding. There were therefore a few genuine leader farmers, and much more could have been made of their successes in order to help address the problems with which the other farmers were failing to cope. Monitoring, evaluation and then feeding the results, conclusions and recommendations of these exercises into the system would fully have justified the 'experiment' of the national leader farmer GMDP scheme. Such work, together with the multiple tasks of maintaining the wider vision of the original Khabura Project could have been - and still could be - the tasks of a Rural Development Centre, as described below.

6. Prospect and retrospect

6.1 Prospect: integration through a Rural Development Centre

The Khabura Project had - and in concept still has - an important future role to play as what could be called a Rural Development Centre (RDC), with roles which would include adaptive research, farm systems development and training. A key factor for the success of the RDC would be competent, well motivated and well supported staff. One of their most important underlying tasks would be to create on-going and fruitful dialogue between the rural communities and those national and international organisations whose decisions affect (or might affect) their work and livelihoods.

The RDC would have to define its research and development roles to be different from those of the MAF Agricultural Research Stations (such as RLRS) on the one hand and the MAF regional extension centres on the other hand. Whilst research of a more fundamental nature, on particular topics, would be undertaken on the research stations, the RDC would in part focus on 'adaptive research' objectively to test technologies and methods in the field. Its primary research task however would be to integrate different aspects of fundamental research to produce coherent farm production systems which not only fitted the current (and likely future) local, physical, economic and human constraints and opportunities being faced by the rural communities but which were also systems that the farmers could, and would in practice, adopt. The proposed relationship between the RDC, the Agricultural Research Station (such as RLRS), extension centres and farmers is illustrated in Figure 36. In addition to taking its fundamental research data from the ARS, the RDC would gain its knowledge of local farm systems from the farmers and other rural producers. Working closely with the farmers it would then suggest modifications to those systems taking current and likely future options and constraints into consideration, not neglecting those affecting the market. These suggestions, presented as farm systems packages, would then be more widely disseminated through the extension centres.

It would also remain important for the RDC to continue working directly with at least a limited number of farmers in order to keep in step with their real world ambitions and limitations on their actions. In practice the Khabura Project, as has been shown, worked with farmers mainly on a one-on-one basis. Initially different farmers were perhaps interested in only a single aspect of the work. However, this led naturally to the creation of a group of about 60 so-called participant farmers who adopted a simplified version of the Project's livestock farm system as a whole. We were moving, again in an evolutionary manner, towards working with groups of farmers, when the Minister closed the Khabura centre in favour of a national GMDP, as has been described. But undoubtedly the experience of working directly with farmers, and other rural producers, was an essential mutual learning experience. Untested ideas became more realistic and, as the Project's understanding of real options and constraints grew, the information was passed on to the Ministry, the Bank and other institutions.

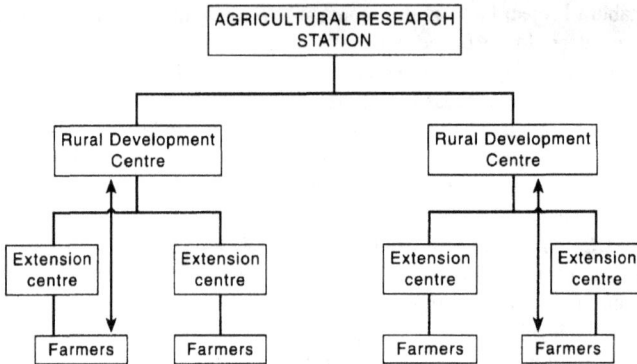

Figure 36. Relationship of the RDC to other key organisations and farmers

In addition to working directly with farmers it would be equally important for the proposed RDC to maintain a trials and demonstration farm on which to introduce new ideas before demonstrating them off-station. The Khabura Project's experience at al-Khabura showed that such a farm has real value if it is:

(a) Within a farming community: on a piece of land in the midst of a group of typical farms similar in size and soil quality, and water availability and quality;

(b) More productive: with yields and productivity increases clearly visible to the naked eye;

(c) Less labour intensive: many of Oman's men in the most active age groups have found employment outside their villages. Therefore new farm systems at Khabura have had to minimise hard manual work and reduce all labour inputs, though this is becoming less true today as a result of rapid population growth and reduced opportunities for labour in the Gulf States.

(d) Not too complicated: most of Oman's small farmers do not have the educational background to follow complicated techniques, nor does the maintenance backup exist at village level to operate such techniques where they involve complex modern equipment. Therefore a farm system must only become complex where this gives an important production or water conservation advantage.

222

(e) Economically viable: the RDC should be run in such a way that it can be seen by the farmers to make money, or at least to be market orientated. Only this will attract serious interest and imitation.

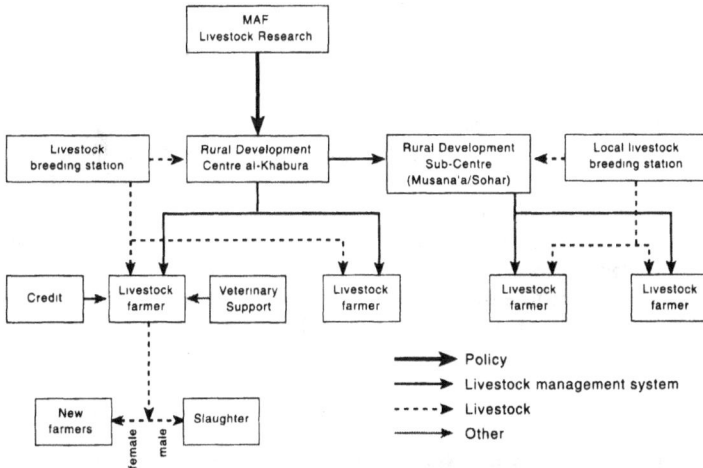

Figure 37. Livestock research and extension system incorporating the RDC and RDC sub-centres at Musana'a and Sohar

The RDC could concern itself with a range of farm systems, including associated rural industries. But as an example based on the livestock related work discussed in this volume, the RDC would work within the context of a network of livestock organisations (Figure 37) which would provide: information about the production system, improved quality breeding stock, veterinary support, credit, technical know-how and appropriate goat and sheep marketing structures. It is important that the senior staff of the RDC should have sufficient authority to be able to call on the livestock organisations at a senior level, and to be able to influence their decisions as a result of what is being learned at grass-roots level. One important means of interaction would be for RDC staff to bring institutional officials and farmers representatives into direct discussion with each other, perhaps with the RDC staff acting as 'interpreters' if it finds that neither side is speaking the other's language. Some relevant organisations and topics for on-going direct discussion with farmers' representatives would include the following:

(a) Oman Bank for Agriculture and Fisheries: For livestock OBAF has two principal roles, both involving financial support for farmers wishing to build up their goat and sheep herds:

223

- Credit is needed by farmers to purchase improved goats and sheep, to buy the materials and labour needed to construct livestock housing, and to pay for essential equipment. The credit should be made available if the farmers, through the RDC, convince OBAF that they are competent and that they will agree to make and use appropriate housing and equipment, and that they will be open to advice and support from RDC staff. The terms, however, on which credit is given would need to be simplified (e.g., on the proof of land ownership) and the whole process of credit issue speeded up.

- For livestock numbers to grow rapidly it is essential that the cross-bred progeny born on a farmer's land, but surplus to his breeding requirements, should be available to other farmers for breeding, and not sold for slaughter. The young stock could be taken back to the RDC, if the farmer needed to remove them from his feed bill, for raising until resale. But to ensure that this happened, a scheme would be needed whereby OBAF bought the young stock from the first farmer at an attractive price and resold to the new farmer at a subsidised price, or with credit.

(b) Marketing organisation, PAMAP: The national livestock marketing policy will need to be under constant review if Oman's new goat and sheep farmers are to have sufficient market confidence to encourage them to increase their herd size. Local production was (and remains in 1997) much below total demand. In these circumstances it is right that cheap sheep should be imported from Australia and elsewhere. However, if the sheep import numbers grow freely they will, eventually, act as a severe disincentive to local producers. Omani stock farmers will not be able to compete and will go out of business. Imports, therefore, will need to be regulated in some way, and balanced against demand so that local production gradually increases. Inevitably, while the local livestock industry is growing this will entail some form of output subsidy. This should be graduated so as to reward quality as well as quantity. Rewarding quality will result in an efficient livestock industry - one which favours the better new stock farmers and discourages the *shawawi* pastoralists who produce too many animals at the expense of the thin natural browse on the plains and mountains but put too little flesh on the bone.

(c) Research: Some problems will require a research input, concerned with both natural resource and socio-economic issues. The RDC staff may be able to address some of these questions themselves. However, they would also need to have the freedom and authority to draw upon research organisations to help, including the research staff and students of Sultan Qaboos University.

Training would be central to the RDC's role. Based on the experience of the Khabura Project this could take many forms. The Project trained employees, farmers, technical specialists and extension centre staff (in ways that have been described) and it tried to create opportunities for some of the trainees to run their own enterprises. These latter included, in addition to goat and sheep farmers, the provision of rotavator, irrigation and paraveterinary services, and workshop activities such as manufacturing channel liners, small-farm equipment and woven goods. In the period 1987/8 to 1988/9 the Khabura Project was also increasingly involved in providing the hands-on elements of training for extension centre staff in courses as diverse as animal husbandry and health, irrigation and forage production, dairying and small-farm management. All of these training roles could be further developed within the remit of the proposed RDC. The emphasis should remain with the practical and the hands-on. This would encourage collaboration with other organisations better equipped to provide the theoretical background, including Sultan Qaboos University and Nizwa College.

Finally, and with more subtlety, the RDC should provide motivation and encouragement. Although an important part of motivation comes through market confidence and economic returns, other forms of motivation and encouragement can also play vital roles. For example, the RDC might establish a Farmers' Association with which it would co-ordinate its activities. Groups of farmers would then be invited to meetings to question visiting specialists, watch audio-visual presentations and take part in demonstrations. Also, an annual regional show would encourage friendly rivalry. Visits to other farms or to other agricultural organisations such as OBAF, MAF research stations and Sultan Qaboos University's Faculty of Agriculture would build information networks and promote discussion.

From the above description of an RDC designed from the Khabura Project's experience of developing small-farm livestock systems, it is possible to generalise more broadly, as below:

An RDC should be regarded as an entity which will focus on a particular combination of location, community and production systems. These need to be chosen with some care. Ideally the combination will be recognised as important in terms of sub-regional needs, and representative of the situation in other parts of the country - or even similar situations in other arid lands. In this way maximum benefit from the RDC experience will accrue not only in the immediate locality but also elsewhere. The rural or small farm production systems under examination will need to be of real significance to the community(ies) they serve. They will also have to be feasible from the standpoints of available resources, both human (time and expertise) and financial. The work should be seen where possible to build on previous experience, either locally or from further afield; creating a sense of continuity and dynamic momentum is an important part of the process. Finally, the production system must be so designed that it can achieve visible progress within the patience span of those who are supposed to be benefiting from it and those responsible for allocating its funds.

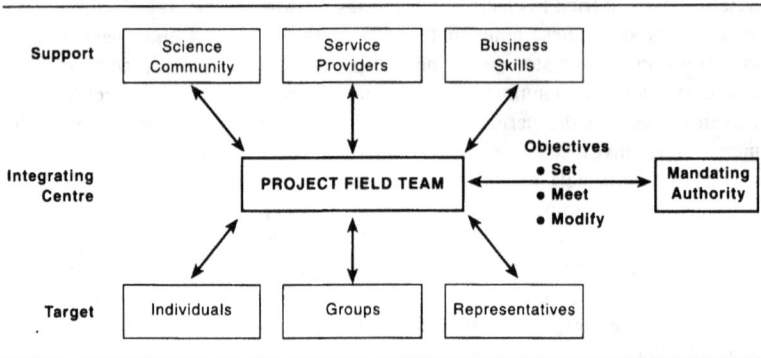

Figure 38. Generalised structure of an RDC

Figure 38 is an example of the general structure of an RDC designed to deliver sustainable benefit for both development and conservation. It deliberately gives central position to the so-called Project Field Team (PFT). This position reflects the need for the PFT to have real authority and, at the same time, to be accountable to all the other RDC elements with which it interacts. It is the integrating centre of the programme and process.

A key initial (and on-going) relationship is with the Mandating Authority. This authority is likely to be the government and will have provided the PFT with its initial objectives and funds. As the bold double-headed arrow in Figure 38 indicates, the two-way relationship between the mandating authority and the PFT is crucial. In concept we are concerned with a fluidly designed programme whose objectives can be changed or modified in the light of experience in the field. We need to move away from the idea of a highly structured and tightly defined short-term project with its narrow terms of reference and its precisely itemised budget allocations, and towards a programme whose remit is clear in terms of its broad objectives but which allows (indeed encourages) room for on-going choice between different means and pathways towards achieving those objectives. There is more likelihood of achieving real influence - as a result of gaining a deep understanding of the local situation - from a longer-term responsive programme with relatively small core funding, than from a shorter-term project (even if better funded) which has a narrow set of fixed tasks to perform, terminating in an abrupt cut-off point.

The PFT membership is also crucial. The team has to have sufficient expertise, status, and natural leadership authority to command the attention of potential providers of Support communities and of the members of the Target community. Because they have to be able to interact with both groups with

226

equal facility, and to speak the very different 'languages' of both with equal fluency, it is far from essential that team members should be from the local population. Indeed it might be easier for them to play a disinterested role between different competing local interest groups if they are not. The team members will also, perhaps above all else, need to be dedicated to the cause - it is not a role for people with a 9-5 bureaucratic mentality. They will also require an appropriate mix of technical expertise. It is also essential that they have excellent communication skills, and as much up-to-date communications technology as possible at their disposal to facilitate the receipt and transmission of information and the promotion of contacts and discussion. They will need to bring together, in the field, in on-going interaction, different combinations of representatives of the target and support communities and the mandating authority. Each will have a lot to learn from the others. The PFT will make mistakes, but this should be expected and be seen as part of the learning process.

The PFT will have three support groups, the Science Community, Service Providers and Business Skills (Figure 38). Perhaps the easiest roles to understand are those of the Service Providers. The local communities, even in the most marginal rural areas, will have a strong perception that the provision of more and better quality health and education services and utilities is desirable. These products can also be delivered in relatively standard and non-controversial packages. Business Skills will be of equal importance. The requirement will be mainly for small enterprise skills, which will help the communities to increase the value added of their local resources and seek to ensure that the benefit accrues to the communities, and not simply to wealthy outsiders who have no innate interest in the area or its people. The encouragement of local entrepreneurial initiatives will also help the project avoid creating a project dependency culture.

Equally significant, though more complicated to integrate into the project, is the role of the Science Community. Any change, if it is to be beneficial in the long-term, will need to be based on new knowledge, much of which will derive from scientific research. But delivering that knowledge can take a long time. The scientific process, particularly when dealing with something as complex as a rural system, can be slow, and the recommendations hedged with qualifications. All too often neither the Mandating Authority nor the Target community have sufficient confidence or patience, or understanding, to allow it to play its proper role. The PFT should be able to help articulate the research tasks to be undertaken that have been identified by the Target community, to ensure farmer participation in the research (Okali *et al.*, 1994), to ensure that the recommendations are matched to the conservation and development requirements of the community, and to create a context which promotes their implementation. Conclusions and recommendations from research and other inputs will need to stem from an interactive process of action and debate with the Target community, convened by the PFT.

Members of the PFT will be in daily contact and communication with the Target community, the people of the locality. Both sides will have a lot to learn, each from the other. Indigenous, local knowledge and new, imported knowledge will need to be woven together to yield a common understanding about what actions will be most beneficial to the community - or sub-sections of it. Almost certainly the starting point will be development, rather than conservation, of scarce arid land resources. People will need to see some prospect of short or medium term gain before they will want to become involved.

Promoting ideas about conservation will be an altogether subtler task. Only if people have confidence in their long-term 'ownership' of the resources in question is it likely that they will be responsive to concepts of conservation and sustainability, and less intensive resource exploitation in the short term. The general task, therefore, is to promote circumstances in which people see themselves as stakeholders in the area. Only then will they be interested in 'appropriate development' and be in a position to consider the demands of conservation in their (and their families') own long-term interests. Creating such circumstances is the role of government, which is another reason why the relationship between the PFT and the Mandating Authority is so important. An RDC, led by a respected PFT, is well placed to facilitate the kind of dialogue between local stakeholders and governmental decision-makers which can lead to necessary changes in the local socio-political and economic context that will allow these stakeholder-friendly circumstances to be brought about. This is not to be confused with government interference leading to a dependency culture. An RDC will make acceptable mistakes, but its unforgivable crime will be to reduce the community members' sense of responsibility for, or freedom of action for, the conservation of their environment or the development of their resources.

It is of fundamental importance that all concerned should see the RDC activities as being experimental. It is very rare for projects which are aimed at changes in the management of sparse rural resources to be fully confident of the outcome. Rural systems are generally far too complex for this to be so. In any case, external factors are likely to change sufficiently markedly during an RDC's lifetime to make modifications of the original goals desirable. But if the work is accepted as experimental, and the uncertainty that this implies is seen as a virtue and not a weakness, then many valuable things follow. First, the PFT can be less prescriptive and more responsive to the local community and local environmental conditions as it develops a deeper understanding of them, gained through direct experience. Second, it is conceptually simpler to make use of the expertise of the research community. Third, it will be easier to define a series of appropriate and changing roles for the providers of services and business skills. Fourth, it will conceptually easier for the mandating authority to accept modifications to its original objectives, and act accordingly. Finally, it will impose a necessary burden on the PFT to force the initial ideas, and all subsequent new ideas, through a cycle of trials, demonstrations, monitoring, evaluation, modification and re-trial, as illustrated in Figure 39. Nobody will then lose face if some of the original ideas fail, or need modifying. Almost all ideas will be improved in the process, and for those that fail - at least it will be known why they fail. It is essential that evaluation should not only be technical but also social, economic and ecological. Within the community perhaps the main criteria for judging an idea to be successful are, first, if the people like it and, second, if they have the interest, the capacity and the resources to take-long term management responsibility for it.

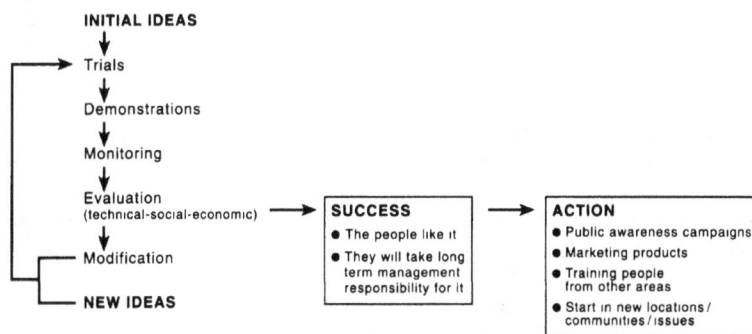

Figure 39. Cycle for testing and implementing production system ideas

If ideas are judged to be successful, this creates the opportunity and the requirement for the PFT and the Mandating Authority to undertake certain additional activities external to the community. Such activities may include a combination of public awareness campaigns and/or assistance with the marketing of products and services in those areas which lie beyond the reasonable reach of the community. They may also include training people from further afield - private citizens and future staff of other PFT's - by exposing them to what has been achieved.

If such an RDC structure is in place it should be able to assist not only with farm systems, such as the livestock systems that have been under discussion in this volume, but also with the wider issues of rural community 're-development' - that re-creation of mutual self-reliance and shared responsibility for the community and its environment which characterised the villages in Oman in pre-oil days.

6.2 Retrospect: evaluating the experience

What did the Khabura Project, including its contributions to ABARC, GMDP and RLRS, achieve? At the outset it was very fortunate, thanks to Shell and

229

PDO sponsorship, to have had the time to learn about the existing, rapidly changing and decaying rural communities in northern Oman. It had also been provided with the financial resources to employ and equip a strong Project Field Team (in the terminology of Figure 38), whose membership evolved over the years to reflect changing requirements. The Project also had a clear focus, based on the development of small-farm livestock production systems, later expanded to include irrigation systems, which were perceived by the community to be important.

The Project was generally successful in its interactions with the Target communities of local producers. Relations between the members of the PFT and local producers were usually good to very good - hard work and effective development of ideas and production practices went hand in hand with real friendship and good humour. In practice we worked mainly with Individuals (Figure 38) but were evolving ways of working with Groups by the time the Project centre at al-Khabura was closed by the Minister of Agriculture in 1989. In every aspect of the work the PFT's ideas and practices evolved and improved as a result of close working relations between staff, farmers and other producers. If initially people were properly cautious about adopting new ideas until they had seen them working in practice, later, as exemplified by the trickle irrigation trials, some farmers were happy to test new ideas on their own land knowing that for the Project's staff as well as themselves they were trials. Some ideas initially passed into the community via employees on the Project's farm who were themselves farmers. If this work had continued, through into the 1990s, ideas and practices would have further evolved and constantly adapted to the changing social and economic context in which the communities lived and worked. The processes indicated in Figure 39 would have continued to the point where public awareness campaigns, product marketing and training people from other areas would have greatly expanded. New centres or sub-centres might have opened - as was often discussed during 1985 in the lead-in to the 1986-90 Plan. Also, Omani field staff would gradually have replaced expatriate staff as the first students of agriculture graduated from SQU and the technical colleges.

However, the Project was less successful in attracting a wide range of effective external Support (Figure 38). The Science Community within Oman was small. SQU opened only in 1986, and in practice the MAF livestock research stations had little clear scientific information to offer. From outside Oman, however, the Project obtained scientific support from a number of sources, principally from the University of Durham which undertook the research surveys on which the Khabura Project was subsequently based. Durham also provided soil science expertise. Newcastle University assisted with studies of animal nutrition, and Edinburgh with aspects of goat and sheep breeding. A number of independent international livestock scientists helped with the selection of the Project's breeds of exotic goats and sheep. Non-university British contributions came mainly in the form of research into irrigation channel technology from ITDG. These contributions are listed in order to illustrate the role that a well-informed PFT can, in practice, perform in securing valuable contributions from social and natural scientists, and technologists, from different parts of the world.

Service Providers of direct relevance to the Project's aims were, locally, few in number. For example, at the outset the Project had to import its own goats, sheep, Rhodes grass seed, rotavators and other goods and materials, though they became much more readily available through the local private sector in later years. To some extent the Project created its own Service Providers, notably the paravets, and it tried as far as possible to use materials available from the al-Khabura *suq*. Potentially one of the most important Service Providers was OBAF, valuable also because of its genuine orientation towards stimulating small private enterprises. In practice with the Project's help the channel workshop secured a loan, and the Bank also showed great interest in supporting the development and use of the Project's low-technology, trickle irrigation system. However, for a variety of reasons we failed to make a sufficiently strong link with OBAF. Partly associated to this lack, the Project did not give enough attention to the important role to be played by Business Skills. In other words, the Project gave insufficient thought, particularly in the early years, to the creation of sustainability through viable rural enterprises (Kinsey, 1987a). Unfortunately the Project was not encouraged in this direction by its links with MAF and MoSAL or its contacts with MCI. MAF was less concerned with the creation of viable farming enterprises than with the provision of highly subsidised input services, which effectively inhibited the rural private sector. MCI's unfortunate contribution to the Khabura Project was to undermine the nascent dairying unit by adopting GCC regulations suited to industrial-scale dairy units without consideration of the needs of small dairy schemes. Meanwhile, MoSAL and MCI between them could not create an effective means of providing small-scale credit to assist the weaving groups at al-Khabura and Ayn Amq. MoSAL, having supported the weaving project over several years to the point where many women were involved in the work, wanted to keep control by - very inefficiently - retaining a role in the marketing process. In order to overcome this lack of business orientation by the ministries the Project would have had to make a major input into raising awareness of its importance and how to set about stimulating an enterprise culture. Unfortunately, the Project, even when it awoke fully to the need, did not have the resources or the remit to make such an input.

However, the major weakness of the Project lay in its relationship with the Mandating Authority (Figure 38), and most particularly with MAF. The relationship between a small field centre and something as large and amorphous as a ministry is bound to be difficult, but for the effective working of a Rural Development Centre, a strong, positive and dynamic relationship is of crucial importance, particularly in setting, meeting and modifying objectives, informing

Ministry policy-making based on field experience and acting on the knowledge so gained. MAF (and MoSAL as far as the spinning and weaving project was concerned) could have assisted the Project's progress in many more ways than they did.

But why in practice were these key relationships with government not more effective, even when from January 1986 all the work was sponsored and financed by the ministries? Certainly it was difficult for MAF up until 1986 to know how to react to the Project - the Ministry was not responsible for it and had had minimal input into the design of its programmes. However, it had, from 1977, directly sponsored and financed two research and development activities, the first phases of the spinning and weaving project and the joint honeybee projects, so the Ministry was generally positive about the Project, as is also shown by its full sponsorship and funding of all the Project's activities from 1986 until 1993/4. In practice the Project's staff were, moreover, in constant contact with a wide range of senior MAF officials and advisers, including regular meetings between the author and the Minister. However, these contacts were not structured into the MAF bureaucratic system. No clear decision-making pathway was ever established. Meetings, even when formally arranged and attended by several officials, often led nowhere even though very positive in tone. For some officials it was preferable to do nothing rather than carry an idea through the Ministry machinery into action. Then, by contrast, key decisions affecting the Project were made by the Minister in person almost entirely on his own judgement.

The Ministry also had little interest in supporting local private enterprise, as stated above, or in the processes of monitoring and evaluating progress and using the information so gained to improve performance. MAF also had no interest in the wider issues of rural development, or in concepts such as encouraging mutual self-reliance or the efficient use of scarce rural resources. There was therefore a 'culture' gap which seemed at times completely unbridgeable.

Furthermore, the government as a whole, with a very laudable desire to help its citizens, found it difficult to strike a proper balance between its wish to provide cheap food and its desire to become agriculturally self-sufficient. Farmers, and other rural producers, struggling to adjust age-old practices to cope with new economic and social realities, were discouraged by cheap imports of livestock and other produce. This made the farmers and craftspeople reluctant to invest in change, and the task of the Project, and indeed the Ministry, much more difficult.

The Minister did not see value in the ideas, outlined above, for a Rural Development Centre at al-Khabura. Indeed, this would have been practically difficult because some of the ideas lay beyond MAF's terms of reference. The Minister's vision for the Project was in one sense much more ambitious, to make it national in scope, but in another sense much narrower, to focus solely on goat and sheep breeding, on dairying and on the 300 'leader farmer' scheme for demonstrating nationally the system which the Project had been developing with its 60 'participant farmers' at al-Khabura.

Paradoxically the Minister's narrower vision made the Rural Development Centre concept more important. If the RDC had come into existence it could now be acting on the results, conclusions and recommendations of the monitoring and evaluation exercises that the Project undertook. It would therefore be seeking the following: to ensure that all the elements of the livestock farm system were in place, and to work directly with both extension centres and farmers in order to maximise the benefits to the rural communities of the investments made. It would also be training technicians, stimulating related local manufacturing, craft-industry and service enterprises and encouraging a growing degree of mutual self-reliance within an improved government policy context.

Khabura Project reports and other publications cited

Addy, B.L. and Taylor, N.M. (1991) The first goat development project: a report on the first three years of a monitoring programme to assess the productivity levels achieved on participating farms. Ministry of Agriculture and Fisheries, Oman.

Addy, B. (1989) A proposal for the development of programmes to support the Goat Development Project and to improve and co-ordinate the training of field staff in the Department of Animal Wealth, Ministry of Agriculture and Fisheries, Oman.

al-Nakib, F.M.S. (1994) 'Animal breeding results', in P.N. Ward (ed) Genetic improvement of Omani sheep and goats, Wadi Quriyat Animal Breeding and Applied Research Centre 1990-93, Department of Livestock Research, Ministry of Agriculture and Fisheries, 101-6.

al-Noaim, A. and Farnworth, J. (1973) The effect of N fertilizer and planting density on the yield and compositions of Rhodes grass under irrigation in Saudi Arabia, Univ. College of North Wales and Ministry of Agriculture and Water, Saudi Arabia, Joint Agriculture and Research Development Project, Publication 12.

Alexander, M.J. (1985) The effect of cultivation and irrigation on the soils of Khabura, Khabura Development Project, Sultanate of Oman, Preliminary reports, 7B(1).

Alexander, M.J. (1990) Further studies on the effect of irrigation and farm management on the soils of Khabura Action Research Centre, Khabura Action Research Centre, Ministry of Agriculture and Fisheries, Sultanate of Oman.

Allen, C.H. (1987) *Oman: the modernization of the Sultanate*, London: Croom Helm.

Barker, A.D.P. (1989) Analysis of feedstuffs grown and used at Khabura, 1983-1989, Khabura Action Research Centre, Ministry of Agriculture and Fisheries, Sultanate of Oman, Preliminary reports, 8D(2).

Barker, A.D.P. (1990) Tools and equipment for the small livestock farmer, Khabura Action Research Centre, Ministry of Agriculture and Fisheries, Sultanate of Oman, Preliminary reports, 8D(2).

Barker, A.D.P. (1991) Farm buildings, Khabura Action Research Centre, Ministry of Agriculture and Fisheries, Sultanate of Oman.

Barnwell, R. (1984a) Goat cross-breeding programme for local farmers in the Khabura area, Khabura Development Project, Sultanate of Oman, Preliminary reports, 1C(1).

Barnwell, R. (1984b) Rhodes grass seed purchase and cultivation by farmers in the Khabura region, 1981-1983, Khabura Development Project, Sultanate of Oman, Preliminary reports, 6C(1).

Barnwell, R. and Massey, R. (1983) Time for treatment in Oman, *International Agricultural Development*, March/April, 10-13.

Bell, G.E. and Dutton, R.W. (1979) A preliminary cost-benefit analysis of running a goat (sheep) small-holding on the Batina, Khabura Development Project, Sultanate of Oman, Preliminary reports, 3(1).

Bertram, G.C.L. (1948) The fisheries of he Sultanate of Oman. Report to the Sultan of Muscat and Oman.

Birks, J.S. (1982) 'The shawawi', in H. Bowen-Jones and R.W. Dutton (eds) Livestock. Research and development surveys in northern Oman, Vol IVb, University of Durham, section 5.9.

Birks, J.S. (1984a) 'Occupations and employment', in H. Bowen-Jones and R.W. Dutton (eds) Demography, employment and migration. Research and development surveys in northern Oman, Vol VIIa, University of Durham, section 8.2.

Birks, J.S. (1984b) 'Movements of migrant labour from part of northern Oman', in H. Bowen-Jones and R.W. Dutton (eds) Demography, employment and migration. Research and development surveys in northern Oman, Vol VIIa, University of Durham., section 8.3.

Birks, J.S. and Letts, S.E. (1976) The Awarmr: specialist well and falaj diggers in northern Oman, *J. Oman Studies*, 2:93-96

Bowen-Jones, H. (1971) Proposal for research and development surveys in northern Oman, Centre for Middle East and Islamic Studies, University of Durham [mimeo].

Bowen-Jones, H. (1974a) Memorandum on a possible rural development project in Oman, Centre for Middle East and Islamic Studies, University of Durham [mimeo].

Bowen-Jones, H. (1974b) Oman - al-Khabourah development project, Centre for Middle East and Islamic Studies, University of Durham [mimeo].

Bowen-Jones, H. and Dutton, R.W. (1984a) Agricultural production and/or rural development, *The Arab Gulf Journal*, 4(1): 51-64.

Brokensha, S. (1980) Khabura spinning and weaving project, annual report 1979/80.

Brokensha, S. (1981) Final report by weaving specialist.

Brown, K. (1988) 'Ecophysiology of Prosopis cineraria in the Wahiba Sands, with reference to its reafforestation potential in Oman', in R.W. Dutton (ed) *J. of Oman Studies*, Special Report No. 3, 257-70.

CAB, (1980) *The nutrient requirements of ruminant livestock*, Commonwealth Agricultural Bureaux.

Clarke, J.I. and Fisher, W.B. (eds) (1972) *Populations of the Middle East and North Africa: a geographical approach*, London: Univ London Press.

Clements, F.A. (1980) *Oman the reborn land*, London and New York: Longman.

Conroy, M.A. (1992a) The goat development project monitoring programme, Ministry of Agriculture and Fisheries, Sultanate of Oman.

Conroy, M.A. (1992b) Goat markets and prices in northern Oman, Ministry of Agriculture and Fisheries, Sultanate of Oman.

Conroy, M.A. (1992c) A financial analysis of intense goat production in northern Oman, Ministry of Agriculture and Fisheries, Sultanate of Oman

Conroy, M.A. (1992d) Briefing document for the seminar on the profitability of goat production on Goat Development Project farms (2 Dec 1992), Ministry of Agriculture and Fisheries, Sultanate of Oman.

Cooke, K.F. and Massey, R.W. (1979a) Livestock diseases observed in the region of al-Khabura, March-June 1979, Durham University Khabura Development Project, Preliminary reports 1A(1).

Cooke, K.F. and Massey, R.W. (1979b) Animal diseases and veterinary work at Khabura, 25.6.79 to 31.8.79, Durham University Khabura Development Project, Preliminary reports 1A(2).

Cooke, K.F. and Massey, R.W. (1980a) Veterinary activity in the Khabura region: Spring 1980, Durham University Khabura Development Project, Preliminary reports 1A(3).

Cooke, K.F. and Massey, R.W. (1980b) Recommendations for an animal husbandry extension programme (an approach to encourage self-reliance amongst Batina farmers), Durham University Khabura Development Project, Findings and recommendations, No 3.

Cooke, K.F. and Massey, R.W. (1981a) Veterinary activity in the Khabura region: July-August 1980, Durham University Khabura Development Project, Preliminary reports 1A(4).

Cooke, K.F. and Massey, R.W. (1981b) Veterinary activity in the Khabura region: September/October 1980, Durham University Khabura Development Project, Preliminary reports 1A(5).

Cooke, K.F. and Massey, R.W. (1981c) Veterinary activity in the Khabura region: November 1980 to March 1981, Durham University Khabura Development Project, Preliminary reports 1A(7).

Cooke, K.F. and Massey, R.W. (1981d) Factors inhibiting the development of small, private livestock enterprises on the Batina, Durham University Khabura Development Project, Preliminary reports 3(2).

Cooke, K.F. and Massey, R.W. (1981e) Traditional Omani approaches to livestock health and husbandry, Durham University Khabura Development Project, Preliminary reports 1A(6).

Crocker, G. with Heath, C. (1988) 'Traditional crafts: products and techniques', in R.W. Dutton (ed) *J. of Oman Studies*, Special Report No. 3, 501-22.

Crocker-Jones, G. (1979) The spinning and weaving project in the Sultanate of Oman, *Warp and Weft*, September issue.

Devendra, C. and Burns, M., (1983) *Goat production in the tropics*, London: Commonwealth Agricultural Bureaux.

Donaldson, W.J. (1978a) 'Fishing and Fish Marketing', in H. Bowen-Jones (ed) Research and development surveys in northern Oman, Vol V, University of Durham.

Donaldson, W.J. (1978b) 'Marketing', in H. Bowen-Jones (ed) Research and development surveys in northern Oman, Vol VI, University of Durham.

Donaldson, W.J. (1981) 'Fisheries in the Arabian peninsula', in J.I. Clarke and H. Bowen-Jones (eds) *Change and Development in the Middle East*, London and New York: Methuen, 189-198.

Dutton, R.W. (1977) Suggested schema for a small-scale wool spinning and weaving development project, at al-Khabura [mimeo].

Dutton, R.W. (1979) The joint spinning and weaving project until March 1981, and beyond, Durham University Khabura Development Project, Findings and recommendations, No 2.

Dutton, R.W. (1980) 'The agricultural potential of Oman', in M. Ziwar-Daftari (ed) *Issues in development: the Arab Gulf states*, MD Research and Services, 170-184.

Dutton, R.W. (1981) 'A rural community development project in Oman', in J.I. Clarke and, H. Bowen-Jones (eds) *Change and Development in the Middle East*, London and New York: Methuen, 199-212.

Dutton, R.W. (1982a) Cultivation using a 2-wheel tractor on the Batina, Findings and recommendations, No 1, Durham University Khabura development project.

Dutton, R.W. (1982b) 'The date palm', in H. Bowen-Jones and R.W. Dutton (eds) Agriculture. Research and development surveys in northern Oman, Vol IVa, University of Durham, section 5.3.

Dutton, R.W. (1982c) 'Alfalfa and other fodder crops', in H. Bowen-Jones and R.W. Dutton (eds) Agriculture. Research and development surveys in northern Oman, Vol IVa, University of Durham, section 5.4.

Dutton, R.W. (1982d) 'Limes on the Batina', in H. Bowen-Jones and R.W. Dutton (eds) (1982a) Agriculture. Research and development surveys in northern Oman, Vol IVa, University of Durham, section 5.5.

Dutton, R.W. (1982e) 'Wheat and minor crops', in H. Bowen-Jones and R.W. Dutton (eds) Agriculture. Research and development surveys in northern Oman, Vol IVa, University of Durham, section 5.6.

Dutton, R.W. (1982f) 'Village and farm livestock on the Batina', in H. Bowen-Jones and R.W. Dutton (eds) Livestock. Research and development surveys in northern Oman, Vol IVb, University of Durham, section 5.8.

Dutton, R.W. (1983a) Handicrafts in Oman and their role in rural community development, *Geoforum* 14(3):341-52.

237

Dutton, R.W. (1983b) Interdependence, independence and rural development in Oman: the experience of the Khabura development project, *J. of Oman Studies*, 6(2): 317-27

Dutton, R.W. (1984) Integrated livestock and irrigation systems: proposal for the action research centre, al-Khabura, 1986-1990, CORD, University of Durham.

Dutton, R.W. (1986a) *Rural development in Oman*, Oman: PDO.

Dutton, R.W. (1986b) 'Agriculture and the future of the aflaj in Oman', in *BRISMES, proceedings of the 1986 international conference on Middle East Studies*, 349-58.

Dutton, R.W. (1987a) 'An explosive rate of change in Khabura, Oman: immigrants fill the labour vacuum', in R. Lawless (ed) *The Middle East village: changing social and economic relations*, London: Croom Helm, 175-93.

Dutton, R.W. (1987b) Oman's pastoralists: settlement, and the provision of irrigated forage, Ministry of Agriculture and Fisheries, Sultanate of Oman.

Dutton, R.W. (1987c) Field trials, demonstrations and extension: a recommended sheep and goat programme for Northern Oman based on the work of the Khabura project, Khabura Action Research Centre, Ministry of Agriculture and Fisheries, Sultanate of Oman.

Dutton, R.W. (1987d) 'Updating agriculture and associated rural enterprises', in B.R. Pridham (ed) *Oman: economic, social and strategic developments*, London: Croom Helm, 94-117

Dutton, R.W. (1991) Forage irrigation rates: flood and sprinkler, Khabura Action Research Centre, Ministry of Agriculture and Fisheries, Sultanate of Oman.

Dutton, R.W. (1994) The potential of paravets in the tropics: experience from al-Khabura, Oman, Khabura Action Research Centre, Ministry of Agriculture and Fisheries, Sultanate of Oman.

Dutton, R.W. and Abdul Baqi, A.M. (eds) (1992) Sprinkler irrigation manual for the goat multiplication project, compiled by Foster, G.N. with contributions from Sekendar, M.A. and Barker, A.D.P.), Khabura Action Research Centre, Ministry of Agriculture and Fisheries, Sultanate of Oman.

Dutton, R.W. and Steele, M. (1983) Analyses of forage crops grown and used at Khabura, 1978-1983, Khabura Development Project, Ministry of Agriculture and Fisheries, Sultanate of Oman, Preliminary reports, 7A(1).

Economides, S. (1983) *Intensive sheep production in the Near East*, FAO, Animal production and health paper, Rome: FAO, 67.

Eickelman, C. (1984) *Women and communities in Oman*. New York: University Press.

El-Dessouky (1986) A proposal for organizing Artificial Insemination in the Sultanate of Oman, FAO report to Ministry of Agriculture and Fisheries, Sultanate of Oman.

FAO (1995) *The farming systems approach to development and appropriate technology generation*, Rome: FAO.

Farnworth, J. and Ruxton, I.B. (1974) The effect of N on the productivity and composition of Rhodes grass grown under irrigated arid zone conditions, Univ. College of North Wales and Ministry of Agriculture and Water, Saudi Arabia, Joint Agriculture and Research Development Project, Publication 37.

Foster, G.N. (1989) The calculation of an irrigation schedule for sprinkler irrigated Rhodes grass in Northern Oman, Khabura Action Research Centre, Ministry of Agriculture and Fisheries, Sultanate of Oman.

Gauldie, A. (1988) In-service training courses for government extension agents, winter 1988-1989.

Gordon, I. (1983) *Controlled breeding in farm animals*, London: Pergamon Press.

Hall, H.T.P. (1982) *Diseases and parasites of livestock in the tropics*, London: Longman.

Heath, C. (1985a) Weaving project, al-Khabura, current situation and future aims, National Community Development Programme, July 1985.

Heath, C. (1985b) Weaving project, Rustaq, current situation and future aims, National Community Development Programme, July 1985.

Heath, C. (1987) Suggested proposal for the Ministry of Social Affairs and Labour's Community Development Weaving Project for 1987/88.

Heath, C. (1996) Hidden currencies: women, weaving and income generation in Oman. Unpublished PhD thesis, East Anglia University, 1996.

Hill, P. (1970) *Studies in rural capitalism in West Africa*, London: CUP.

Hillman, F. (1983) Fibre-reinforced cement pre-fabricated irrigation channels: progress May-November 1983, Khabura Development Project, Ministry of Agriculture and Fisheries, Sultanate of Oman, Preliminary reports, 8A(3).

Hillman, F. (1984a) Trickle irrigation: a low-cost, low pressure, low technology system, Khabura Development Project, Ministry of Agriculture and Fisheries, Sultanate of Oman, Preliminary reports, 8A(1).

Hillman, F. (1984b) The use of rotavators in Batina gardens and falaj villages, Durham university, Khabura development project.

Hillman, F. (1985) Fibre-reinforced cement pre-fabricated irrigation channels: position in October 1985, Khabura Development Project, Ministry of Agriculture and Fisheries, Sultanate of Oman, Preliminary reports, 8A(5).

Hillman, F. (1986a) Irrigation systems developed for small-farmer requirements, Khabura Development Project, Ministry of Agriculture and Fisheries, Sultanate of Oman, Preliminary reports, 8A(6).

Hillman, F. (1986b) Pre-fabricated fibre-reinforced cement irrigation channels, *Waterlines* 4(4): 22-5.

Horton, L. (1984a) Goat dairy project, Durham University Khabura Development Project, Sultanate of Oman, Preliminary reports, 9(1).

Horton, L. (1984b) Milk yields from cows in the Khabura region, Durham University Khabura Development Project, Sultanate of Oman, Preliminary reports, 9(2).

Horton, L. (1984c) Report on milk project 1983-1984, Durham University Khabura Development Project, Sultanate of Oman, Preliminary reports, 9(3).

Horton, L. (1987) Dairy section final report 1985-1986, Khabura Action Research Centre, Ministry of Agriculture and Fisheries, Sultanate of Oman, Preliminary reports, 9(4).

Hutchinson, J. (1966) Introduction, The transformation of rural communities; World Land Use Survey, Occ. Papers No.7.

ICAE (1987) *Latest developments in livestock housing*, Seminar of the 2nd technical section of the CIGR, International Commission of Agricultural Engineering, University of Illinois, June 22-26.

Kay, M. (1986) *Surface irrigation: systems and practice*, Cranfield Press.

Khamfar, K.A. (1987) Goat breeding farms, subsidy and development project, Ministry of Agriculture and Fisheries, Sultanate of Oman, Aug 1987.

Kinsey, B.H. (1987a) *Agribusiness and rural enterprise*, London: Croom Helm.

Kinsey, B.H. (1987b) *Creating rural employment*, London: Croom Helm.

Kwantes, L.J. (1993) Survey of veterinary activities on goat and sheep development project farms during 1992, Centre for Overseas Research and Development, University of Durham.

Kwantes, L.J. (1994) 'Animal health', in P.N. Ward (ed) Genetic improvement of Omani sheep and goats, Wadi Quriyat Animal Breeding and Applied Research Centre 1990-3, Department of Livestock Research, Ministry of Agriculture and Fisheries, 31-100.

Lancaster, W. (1988) 'Fishing and coastal communities: indigenous economies, decline or renewal', in R.W. Dutton (ed) *J. of Oman Studies*, Special Report No. 3, 485-94.

Letts, S. (1978a) 'Fluctuations in water levels and aquifer yields', in H. Bowen-Jones (ed) Water, Research and development surveys in northern Oman, Vol II, University of Durham, 29-35.

Letts, S. (1978b) 'Wells and the development of pump wells', in H. Bowen-Jones (ed) Water, Research and development surveys in northern Oman, Vol II, University of Durham, 66-126.

Letts, S. (1978c) 'The falaj', in H. Bowen-Jones (ed) Water, Research and development surveys in northern Oman, Vol II, University of Durham, 127-174.

Letts, S. (1982) 'Irrigation practice', in H. Bowen-Jones and R.W. Dutton (eds) Agriculture. Research and development surveys in northern Oman, Vol IVa, University of Durham, section 5.2.

Lochhead, A. (1983) Monthly report 5; May 1983 - Community Development Weaving Project, Ministry of Social Affairs and Labour, Sultanate of Oman.

Lochhead, A. (1984a) Weaving project, final report, National Community Development Programme, Ministry of Social Affairs and Labour, Sultanate of Oman.

Lochhead, A. (1984b) Handover notes, Community Development Weaving Project, Ministry of Social Affairs and Labour, Sultanate of Oman.

Miller, D.R. (1991) *Economic development planning in the Sultanate of Oman*, Muscat: United Media Services.

Mohammadein, A.D. (1989) Report on development of artificial insemination services in cattle in the Sultanate of Oman.

OBAF (1982) Sheep/Goat Rearing Program: Project 3/82/4, Feasibility Study and Proposal, The Oman Bank for Agriculture and Fisheries, SAO, June 19th.

Okali, C., Sumberg, J. and Farrington, J. (1994) *Farmer participatory research: rhetoric and reality*, London: Intermediate Technology Publications.

Penrose (1971) 'Oil and state in Arabia', in D. Hopwood (ed) *The Arabian peninsula: society and politics*, 271-85.

Rogerson, I. (1989) Report on veterinary and animal husbandry phase 1 courses held at KARC.

Rogerson, K. and Omer Binofuf, M.H. (1990) A survey to determine the potential for dairying and interest in A.I. services in the South Batina area, Rumais Livestock Research Station, Department of Animal Wealth, Ministry of Agriculture and Fisheries, Sultanate of Oman.

Schultz, T.W. (1964) *Transforming traditional agriculture*, Yale Univ. Press.

Schultz, T.W. (1972a) 'Production opportunities in Asian agriculture: an economist's agenda', in W.L. Johnson and D.R. Kamerschen (eds) *Readings in economic development*, Cincinnati, South Western Publishing Co.

Schultz, T.W. (1972b) 'Investment in human capital in poor countries', in *Readings in human development*, Cincinnati: South Western Publishing Co.

Schumacher, E.F., (1973) *Small is beautiful: a study of economics as if people mattered*, London: Blond & Briggs.

Sherwood, H.S. & A. (1984) Lambing results of synchronised breeding of ewe lambs, January 1983-June 1983, Khabura Development Project, Sultanate of Oman, Preliminary reports, 1B(6).

Sidgwick, G.N.B. (1994) 'Farm management', in P.N. Ward (ed) Genetic improvement of Omani sheep and goats, Wadi Quriyat Animal Breeding and Applied Research Centre 1990-3, Department of Livestock Research, Ministry of Agriculture and Fisheries, 15-30.

Skeet, I. (1992) *Oman: politics and development*, London: Macmillan.

Steele, M. (1983a) Hormonal synchronisation of breeding in goats, Khabura Development Project, Sultanate of Oman, Preliminary reports, 1B(2).

Steele, M. (1983b) Initial results of sheep cross-breeding programme: Chios x Omani, Khabura Development Project, Sultanate of Oman, Preliminary reports, 1B(4).

Steele, M. (1983c) Synchronised sheep breeding trials at Khabura, November to May 1981/82, and September to February 1982/83, Khabura Development Project, Sultanate of Oman, Preliminary reports, 1B(5).

Steele, M. (1983d) Preliminary field trials with new forage legumes, Khabura Development Project, Sultanate of Oman, Preliminary reports, 6A(2).

Steele, M. (1983e) Rhodes grass rotations, Khabura Development Project, Sultanate of Oman, Preliminary reports, 6B(2).

Steele, M. (1983f) The production, use and sale of manure/compost from sheep/goat waste at Khabura, Durham University Khabura Development Project, Sultanate of Oman, Preliminary reports, 5A(1).

Steele, M. and Dutton, R.W. (1983a) Alfalfa production, April 1981-March 1983, Khabura Development Project, Sultanate of Oman, Preliminary reports, 6B(1).

Steele, M. and Dutton, R.W. (1983b) Rhodes grass yields, 1981-1983, Khabura Development Project, Sultanate of Oman, Preliminary reports, 6A(1).

Stephens, M. (1992) Irrigation and fodder production systems: observations made during visits to 99 goat development project farms, Centre for Overseas Research and Development, University of Durham.

Stephens, M. (1993a) Analysis of irrigation systems used for Rhodes grass production: using data collected from all phase 1 goat development project farms, Centre for Overseas Research and Development, University of Durham.

Stephens, M. (1993b) An evaluation of Rhodes grass production under sprinkler irrigation on a sample of goat development project farms, Centre for Overseas Research and Development, University of Durham.

Stern, P. (1979) *Small-scale irrigation: a manual of low-cost water technology*, London: IT Publications.

Taylor, N.M. (1990a) Goats at KARC farm (1977 to 1989), Khabura Action Research Centre, Ministry of Agriculture and Fisheries, Sultanate of Oman.

Taylor, N.M. (1990b) Sheep at KARC farm (1977 to 1989), Khabura Action Research Centre, Ministry of Agriculture and Fisheries, Sultanate of Oman.

Taylor, N.M. (1991a) Goats and sheep at KARC farm (1977 to 1989), Khabura Action Research Centre, Ministry of Agriculture and Fisheries, Sultanate of Oman.

Taylor, N.M. (1991b) Monitoring system for the goat multiplication and development project and sheep multiplication and development project: description and operating instructions, Rumais livestock research station, Ministry of Agriculture and Fisheries, Sultanate of Oman.

Taylor, N.M. (1991c) Goat development project: monitoring report on data logged up to March 1991, Rumais livestock research station, Ministry of Agriculture and Fisheries, Sultanate of Oman.

Taylor, N.M. (1991d) Goat development project: monitoring report on data logged up to 31/7/1991, Rumais livestock research station, Ministry of Agriculture and Fisheries, Sultanate of Oman (data plus notes).

Taylor, N.M. (1991e) An economic analysis of goat and sheep production at the Khabura Action Research Centre farm - 1987 and 1988.

Tiffen, M (1976) *The enterprising peasant: economic development in Gombe Emirate, North-Eastern State, Nigeria, 1900-1968*, London: HMSO.

Ward, P.N. (1994a) Provision of artificial insemination services for cattle in the Sultanate of Oman: artificial insemination project, Rumais 1990-1993, Livestock Research Centres Management Project, Department of Animal Wealth, Ministry of Agriculture and Fisheries, Sultanate of Oman.

Ward, P.N. (ed) (1994b) Genetic improvement of Omani sheep and goats, Wadi Quriyat Animal Breeding and Applied Research Centre 1990-93, Department of Livestock Research, Ministry of Agriculture and Fisheries, 31-100.

Weber, N. (1989) The production of Rhodes grass and its utilisation by goats at the Desert Agriculture Project, Marmul, Petroleum Development Oman, Sultanate of Oman.

Webster, R. (1988) 'Bedouin of the Wahiba Sands, pastoral economy and society', in R.W. Dutton (ed) *J. of Oman Studies*, Special Report No. 3, 461-72.

Wilkinson, J.C. (1977) *Water and tribal settlement in south-east Arabia*, Oxford: Clarendon Press.

Winstanley, G. (1983) Fibre-reinforced cement irrigation channels: progress 1981 to March 1933.

Winstanley, G. and Hillman, F. (1984) Design and cost of sheep/goat pens, Durham University Khabura Development Project, Sultanate of Oman, Preliminary reports 8C(1).

For Product Safety Concerns and Information please contact our EU
representative GPSR@taylorandfrancis.com
Taylor & Francis Verlag GmbH, Kaufingerstraße 24, 80331 München, Germany